Math Mammoth
Grade 8-A Worktext
Canadian Version

By Maria Miller

Contents

Chapter 3: Linear Equations

Chapter 4: Introduction to Functions

Foreword

Math Mammoth Grade 8 comprises a complete math curriculum for the eighth grade mathematics studies. The curriculum meets the Common Core standards.

In 8th grade, students spend the majority of the time with algebraic topics, such as linear equations, functions, and systems of equations. The other major topics are geometry and statistics.

The main areas of study in Math Mammoth Grade 8 are:

- Exponents laws and scientific notation
- Square roots, cube roots, and irrational numbers
- Geometry: congruent transformations, dilations, angle relationships, volume of certain solids, and the Pythagorean Theorem
- Solving and graphing linear equations;
- Introduction to functions;
- Systems of linear equations;
- Scatter plots/bivariate data.

We start with a study of exponent laws, using both numerical and algebraic expressions. The first chapter also covers scientific notation (both with large and small numbers), significant digits, and calculations with numbers given in scientific notations.

In chapter 2, students learn about geometric transformations (translations, reflections, rotations, dilations), common angle relationships, and volume of prisms, cylinders, spheres, and cones.

Next, in chapter 3, our focus is on linear equations. Students both review and learn more about solving linear equations, including equations whose solutions require the usage of the distributive property and equations where the variable is on both sides.

Chapter 4 presents an introduction to functions. Students construct functions to model linear relationships, learn to use the rate of change and initial value of the function, and they describe functions qualitatively based on their graphs.

In part 8-B, students graph linear equations, learn about irrational numbers and the Pythagorean Theorem, solve systems of linear equations, and investigate patterns of association in bivariate data (scatter plots).

I heartily recommend that you read the full user guide in the following pages.

I wish you success in teaching math!

Maria Miller, the author

User Guide

Note: You can also find the information that follows online, at https://www.mathmammoth.com/userguides/ .

Basic principles in using Math Mammoth Complete Curriculum

Math Mammoth is mastery-based, which means it concentrates on a few major topics at a time, in order to study them in depth. The two books (parts A and B) are like a "framework", but you still have some liberty in planning your student's studies. In eighth grade, chapters 2 (geometry), 3 (linear equations) and chapter 4 (functions) should be studied before chapter 5 (graphing linear equations). Also, chapters 3, 4, and 5 should be studied before chapter 7 (systems of linear equations) and before chapter 8 (statistics). However, you still have some flexibility in scheduling the various chapters.

Math Mammoth is not a scripted curriculum. In other words, it is not spelling out in exact detail what the teacher is to do or say. Instead, Math Mammoth gives you, the teacher, various tools for teaching:

- **The two student worktexts** (parts A and B) contain all the lesson material and exercises. They include the explanations of the concepts (the teaching part) in blue boxes. The worktexts also contain some advice for the teacher in the "Introduction" of each chapter.

  The teacher can read the teaching part of each lesson before the lesson, or read and study it together with the student in the lesson, or let the student read and study on his own. If you are a classroom teacher, you can copy the examples from the "blue teaching boxes" to the board and go through them on the board.

- Don't automatically assign all the exercises. Use your judgement, trying to assign just enough for your student's needs. You can use the skipped exercises later for review. For most students, I recommend to start out by assigning about half of the available exercises. Adjust as necessary.

- For each chapter, there is a **link list to various free online games** and activities. These games can be used to supplement the math lessons, for learning math facts, or just for some fun. Each chapter introduction (in the student worktext) contains a link to the list corresponding to that chapter.

- The student books contain some **mixed review lessons**, and the curriculum also provides you with additional **cumulative review lessons**.

- There is a **chapter test** for each chapter of the curriculum, and a comprehensive end-of-year test.

- You can use the free online exercises at https://www.mathmammoth.com/practice/
  This is an expanding section of the site, so check often to see what new topics we are adding to it!

- And there are answer keys to everything.

How to get started

Have ready the first lesson from the student worktext. Go over the first teaching part (within the blue boxes) together with your student. Go through a few of the first exercises together, and then assign some problems for the student to do on their own.

Repeat this if the lesson has other blue teaching boxes.

Many students can eventually study the lessons completely on their own — the curriculum becomes self-teaching. However, students definitely vary in how much they need someone to be there to actually teach them.

Pacing the curriculum

Each chapter introduction contains a suggested pacing guide for that chapter. You will see a summary on the right. (This summary does not include time for optional tests.)

Most lessons are 3 or 4 pages long, intended for one day. Some lessons are 5 pages and can be covered in two days.

It can also be helpful to calculate a general guideline as to how many pages per week the student should cover in order to go through the curriculum in one school year.

Worktext 8-A	
Chapter 1	13 days
Chapter 2	27 days
Chapter 3	21 days
Chapter 4	14 days
TOTAL	**75 days**

The table below lists how many pages there are for the student to finish in this particular grade level, and gives you a guideline for how many pages per day to finish, assuming a 180-day (36-week) school year. The page count in the table below *includes* the optional lessons.

Example:

Grade level	School days	Days for tests and reviews	Lesson pages	Days for the student book	Pages to study per day	Pages to study per week
8-A	90	8	204	82	2.5	12.5
8-B	90	8	182	82	2.2	11
Grade 8 total	180	16	386	164	2.4	12

The table below is for you to fill in. Allow several days for tests and additional review before tests — I suggest at least twice the number of chapters in the curriculum. Then, to get a count of "pages to study per day", **divide the number of lesson pages by the number of days for the student book**. Lastly, multiply this number by 5 to get the approximate page count to cover in a week.

Grade level	Number of school days	Days for tests and reviews	Lesson pages	Days for the student book	Pages to study per day	Pages to study per week
8-A			204			
8-B			182			
Grade 8 total			386			

Now, something important. Whenever the curriculum has lots of similar practice problems (a large set of problems), feel free to **only assign 1/2 or 2/3 of those problems**. If your student gets it with less amount of exercises, then that is perfect! If not, you can always assign the rest of the problems for some other day. In fact, you could even use these unassigned problems the next week or next month for some additional review.

In general, 8th graders might spend 45-75 minutes a day on math. If your student finds math enjoyable, they can of course spend more time with it! However, it is not good to drag out the lessons on a regular basis, because that can then affect the student's attitude towards math.

Using tests

For each chapter, there is a **chapter test**, which can be administered right after studying the chapter. **The tests are optional.** The main reason for the tests is for diagnostic purposes, and for record keeping. These tests are not aligned or matched to any standards.

In the digital version of the curriculum, the tests are provided both as PDF files and as html files. Normally, you would use the PDF files. The html files are included so you can edit them (in a word processor such as Word or LibreOffice), in case you want your student to take the test a second time. Remember to save the edited file under a different file name, or you will lose the original.

The end-of-year test is best administered as a diagnostic or assessment test, which will tell you how well the student remembers and has mastered the mathematics content of the entire grade level.

Using cumulative reviews and the worksheet maker

The student books contain mixed review lessons which review concepts from earlier chapters. The curriculum also comes with additional cumulative review lessons, which are just like the mixed review lessons in the student books, with a mix of problems covering various topics. These are found in their own folder in the digital version, and in the Tests & Cumulative Reviews book in the print version.

The cumulative reviews are optional; use them as needed. They are named indicating which chapters of the main curriculum the problems in the review come from. For example, "Cumulative Review, Chapter 4" includes problems that cover topics from chapters 1-4.

Both the mixed and cumulative reviews allow you to spot areas that the student has not grasped well or has forgotten. When you find such a topic or concept, you have several options:

1. Check for any online games and resources in the Introduction part of the particular chapter in which this topic or concept was taught.

2. If you have the digital version, you could reprint the lesson from the student worktext, and have the student restudy that.

3. Perhaps you only assigned 1/2 or 2/3 of the exercise sets in the student book at first, and can now use the remaining exercises.

4. Check if our online practice area at https://www.mathmammoth.com/practice/ has something for that topic.

5. Khan Academy has free online exercises, articles, and videos for most any math topic imaginable.

Concerning challenging word problems and puzzles

While this is not absolutely necessary, I heartily recommend supplementing Math Mammoth with challenging word problems and puzzles. You could do that once a month, for example, or more often if the student enjoys it.

The goal of challenging story problems and puzzles is to **develop the student's logical and abstract thinking and mental discipline**. I recommend starting these in fourth grade, at the latest. Then, students are able to read the problems on their own and have developed mathematical knowledge in many different areas. Of course I am not discouraging students from doing such in earlier grades, either.

Math Mammoth curriculum contains lots of word problems, and they are usually multi-step problems. Several of the lessons utilize a bar model for solving problems. Even so, the problems I have created are usually tied to a specific concept or concepts. I feel students can benefit from solving problems and puzzles that require them to think "out of the box" or are just different from the ones I have written.

I recommend you use the free Math Stars problem-solving newsletters as one of the main resources for puzzles and challenging problems:

Math Stars Problem Solving Newsletter (grades 1-8)
https://www.homeschoolmath.net/teaching/math-stars.php

I have also compiled a list of other resources for problem solving practice, which you can access at this link:

https://l.mathmammoth.com/challengingproblems

Another idea: you can find puzzles online by searching for "brain puzzles for kids," "logic puzzles for kids" or "brain teasers for kids."

Frequently asked questions and contacting us

If you have more questions, please first check the FAQ at https://www.mathmammoth.com/faq-lightblue

If the FAQ does not cover your question, you can then contact us using the contact form at the Math Mammoth.com website.

Chapter 1: Exponents and Scientific Notation
Introduction

The first chapter of Math Mammoth Grade 8 starts out with a study of basic exponent laws and scientific notation.

We begin with a review of the concept of an exponent and of the order of operations. The next lesson first review multiplication of integers, and then focuses on powers with negative bases, such as $(-5)^3$.

Then we get to the "meat" of the chapter: the various laws of exponents. The first lesson on that topic allows students to explore and to find for themselves the product law and the quotient law of exponents. After that, students find out the logical way to define negative and zero exponent by looking at patterns. They practise simplifying various expressions with exponents, both with numerical values and with variables.

The lesson "More on Negative Exponents" focuses on expressions with a negative exponent in the numerator, such as $7/(a^{-4})$. This is to prepare students for calculations that ask them to find how many times bigger one number is than another, when the numbers are written in scientific notation.

Next, in the lesson "Laws of Exponents, Part 2", students practise applying the power of a power law: $(a^n)^m = a^{nm}$.

Then the chapter has one more lesson on the laws of exponents ("Laws of Exponents, Part 3"), which summarizes the laws and gives more practice. This lesson is not absolutely essential if you're following Common Core Standards. It is presented here to give a summary, to give practice on all exponent laws, including the power of a quotient law which was not dealt with a lot in the previous lessons. This lesson also allows the book to meet the Florida B.E.S.T. standards for 8th grade.

Then we turn our attention to scientific notation, first learning how it is used with large numbers and then with small numbers. The lesson on significant digits follows, helping students to know how to round final answers in calculations with measurements.

The last topic of the chapter is calculations with numbers given in scientific notations. These calculations, naturally, involve many scientific topics such as the atomic world or astronomy.

Pacing Suggestion for Chapter 1

This table does not include the chapter test as it is found in a different book (or file).
Please add one day to the pacing if you use the test.

The Lessons in Chapter 1	page	span	suggested pacing	your pacing
Powers and the Order of Operations	13	*3 pages*	1 day	
Powers with Negative Bases	16	*3 pages*	1 day	
Laws of Exponents, Part 1	19	*3 pages*	1 day	
Zero and Negative Exponents	22	*3 pages*	1 day	
More on Negative Exponents	25	*2 pages*	1 day	
Laws of Exponents, Part 2	27	*3 pages*	1 day	
Laws of Exponents, Part 3	30	*2 pages*	1 day	

The Lessons in Chapter 1	page	span	suggested pacing	your pacing
Scientific Notation: Large Numbers	32	*3 pages*	1 day	
Scientific Notation: Small Numbers	35	*2 pages*	1 day	
Significant Digits	37	*3 pages*	1 day	
Using Scientific Notation in Calculations, Part 1	40	*3 pages*	1 day	
Using Scientific Notation in Calculations, Part 2	43	*3 pages*	1 day	
Chapter 1 Review	46	*2 pages*	1 day	
Chapter 1 Test (optional)				
TOTALS		*35 pages*	13 days	

Helpful Resources on the Internet

We have compiled a list of Internet resources that match the topics in this chapter, including pages that offer:

- **online practice** for concepts;
- online **games**, or occasionally, printable games;
- **animations** and interactive **illustrations** of math concepts;
- **articles** that teach a math concept.

We heartily recommend you take a look! Many of our customers love using these resources to supplement the bookwork. You can use these resources as you see fit for extra practice, to illustrate a concept better and even just for some fun. Enjoy!

https://l.mathmammoth.com/gr8ch1

Scan me

Powers and the Order of Operations

You will recall that we use **exponents** as a shorthand for writing repeated multiplications by the same number. For example, $7 \cdot 7 \cdot 7 \cdot 7 \cdot 7$ is written 7^5.

The tiny raised number is called the **exponent**. It tells us how many times the **base** number is multiplied by itself.

exponent

$$12^{\textcircled{4}} = 12 \cdot 12 \cdot 12 \cdot 12$$
$$= 20\ 736$$

base

The entire expression, 7^5, is a **power.** We read it as "seven to the fifth power," "seven to the fifth," or "seven raised to the fifth power." Similarly, 0.5^8 is read as "five tenths to the eighth power" or "zero point five to the eighth."

The "powers of 8" are the various expressions where 8 is raised to some power: for example, 8^3, 8^4, 8^{45}, and 8^{99} are powers of 8.

The expression 9^1 equals simply 9. In general, $a^1 = a$.

Powers of 2 are usually read as something "**squared.**" For example, 11^2 is read as "eleven squared." That is because it gives us the <u>area of a square</u> with sides 11 units long.

Similarly, if the exponent is 3, the expression is usually read using the word "**cubed.**" For example, 1.5^3 is read as "one point five cubed" because it is the <u>volume of a cube</u> with edges 1.5 units long.

A calculator is not needed for the exercises of this lesson.

1. Evaluate.

   a. four cubed   **b.** 2^4 **c.** 5^3

   d. 0.2^3 **e.** 1^{60} **f.** 100 squared

2. **a.** Which is more, 4^2 or 2^4?   **b.** Which is more, 2^5 or 5^2?

3. Complete the patterns.

a.	b.	c.
$10^1 =$	$2^1 =$	$0.1^1 =$
$10^2 =$	$2^2 =$	$0.1^2 =$
$10^3 =$	$2^3 =$	$0.1^3 =$
$10^4 =$	$2^4 =$	$0.1^4 =$
$10^5 =$	$2^5 =$	$0.1^5 =$
$10^6 =$	$2^6 =$	$0.1^6 =$
$10^7 =$	$2^7 =$	$0.1^7 =$

The order of operations dictates that powers (expressions with exponents) are solved before multiplication, division, addition, and subtraction.

Example 1. Find the value of $5 \cdot 0.1^3 + 0.2^2$.

First the powers: $0.1^3 = 0.1 \cdot 0.1 \cdot 0.1 = 0.001$, and $0.2^2 = 0.2 \cdot 0.2 = 0.04$.

The expression becomes
$5 \cdot 0.001 + 0.04 = 0.005 + 0.04 = \underline{0.045}$.

The Order of Operations (BEMDAS)
("*Best Excuse My Dear Aunt Sally*")

1) Solve what is within brackets (**B**).

2) Solve exponents (**E**).

3) Solve multiplication (**M**) and division (**D**) from left to right.

4) Solve addition (**A**) and subtraction (**S**) from left to right.

4. Find the value of the expressions.

a. $4 \cdot 10^3 - 5 \cdot 10^2$	**b.** $4(5^2 - 2^3)$	**c.** $\dfrac{3}{1^8} + \dfrac{5}{3^2}$
d. $7 \cdot 10^3 - 5(800 - 10^2)$	**e.** $500 - \dfrac{3 \cdot 8}{2^3} + 2 \cdot 8^2$	**f.** $\dfrac{2 \cdot 17 + 2^4}{7 \cdot 7 - 3^2} + 20$

5. Find the value of the expressions.

a. $0.5^2 - 0.2^2 - 0.1^2$	**b.** $3(0.1^2 - 0.2^3)$	**c.** $0.6^2 + 2(1 - 0.3^2)$

6. The table on the right shows a list of powers of 4.

   a. Find the value of 4^7 using the value for 4^6. (Do not use a calculator.)

   b. Which power of 4 is equal to 65 536? Use estimation and the table, not a calculator.

   c. Use the table to check whether $4^2 + 4^3 = 4^5$.

   d. Use the table to check whether $4^2 \cdot 4^3 = 4^5$.

$4^1 = 4$
$4^2 = 16$
$4^3 = 64$
$4^4 = 256$
$4^5 = 1024$
$4^6 = 4096$

7. **a.** Find a power of 3 that is greater than seven squared.

   b. Find a power of 5 that is greater than ten cubed.

   c. Find a power of 1 that is greater than three squared.

8. **a.** If $3^6 = 729$, find the value of 3^8.

  b. If $2^8 = 256$, find the value of 2^{11}.

9. Find the missing exponents.

  a. $10^4 = 100$▮

  b. $2^6 = 4$▮

  c. $9^2 = 3$▮

  d. $0 = 0$▮

  e. 0.1▮$ = 0.0001$

  f. 0.2▮$ = 0.00032$

  g. $625 = 5$▮

  h. $128 = 2$▮

10. Find the value of these powers.

a. $\left(\dfrac{1}{6}\right)^2 =$	**b.** $\left(\dfrac{3}{10}\right)^3 =$	**c.** $\left(\dfrac{2}{3}\right)^4 =$	**d.** $\left(\dfrac{3}{4}\right)^3 =$

Example 2. Simplify $3 \cdot s \cdot s \cdot s \cdot 3 \cdot t \cdot s \cdot t \cdot t$.

We can multiply in any order, so let's reorganise the expression as $3 \cdot 3 \cdot s \cdot s \cdot s \cdot s \cdot t \cdot t \cdot t$.

The variable s is multiplied by itself four times, and t three times. Naturally, $3 \cdot 3$ is 9.

So, we get $3 \cdot 3 \cdot s \cdot s \cdot s \cdot s \cdot t \cdot t \cdot t = \mathbf{9s^4 t^3}$.

---

11. Simplify.

a. $2 \cdot x \cdot x \cdot x \cdot x \cdot x \cdot 7$	**b.** $4 \cdot x \cdot x \cdot x \cdot y \cdot y \cdot 9 \cdot x \cdot y \cdot x$
c. $5 \cdot a \cdot b \cdot b \cdot a \cdot a \cdot 2 \cdot b \cdot 6$	**d.** $0.3 \cdot p \cdot r \cdot p \cdot r \cdot r \cdot 0.2 \cdot r \cdot 10$

12. **a.** Find the value of the expression $10a^4b^2$ when $a = 2$ and $b = 3$.

  b. Find the value of the expression $14x^3y^5$ when $x = 2$ and $y = 0$.

13. When you fold a sheet of paper in half, its area is now only 1/2 of the area of the original paper. Let's say you repeat this process, and fold that paper again in half, and again, and again. How many times do you need to fold a sheet of paper in order for the area of the folded piece to be 1/64 of the area of the original?

Puzzle Corner What is the simple value of $\dfrac{9^6}{9^5}$? There is no need for actual calculations!

Powers with Negative Bases

Review. To multiply and divide integers, simply multiply and divide the absolute values of the numbers, and then determine the sign of the final product. For two integers, the sign is given by this chart (applies also to division):

(negative) · (positive) = (negative)

(positive) · (negative) = (negative)

(positive) · (positive) = (positive)

(negative) · (negative) = (positive)

We will look at the situation of three or more integers in the exercises.

Example 1.

$-8 \cdot 3 = -24$

$7 \cdot (-2) = -14$

$-6 \cdot (-6) = 36$

$5 \div (-5) = -1$

$\dfrac{-40}{-2} = 20$

1. Multiply and divide.

a. $-2 \cdot (-7) =$	**b.** $-72 \div 8 =$	**c.** $7 \cdot (-8) =$	**d.** $54 \div (-9) =$
e. $\dfrac{12}{-3} =$	**f.** $9 \cdot (-9) =$	**g.** $(-7) \cdot (-4) =$	**h.** $\dfrac{-36}{-4} =$

2. Use repeated addition to explain why a product of a positive and a negative integer must be negative.

To multiply three or more integers, you can multiply any two of them first. Then multiply their product by the third integer, and so on.

Example 2. Solve $-5 \cdot 3 \cdot (-6) \cdot 2$.

We can first multiply -5 and -6 to get 30. The problem now becomes $30 \cdot 3 \cdot 2 = 180$

3. Multiply. Note that you can multiply in any order.

a. $(-2) \cdot (-7) \cdot (-1) =$	**b.** $5 \cdot (-2) \cdot (-2) =$	**c.** $(-1) \cdot 6 \cdot 5 =$
d. $(-1) \cdot (-2) \cdot (-3) \cdot (-2) =$	**e.** $(-2) \cdot 2 \cdot (-5) \cdot (-1) =$	**f.** $(-2) \cdot (-1) \cdot (-3) \cdot 5 \cdot (-2) =$
g. $6 \cdot \dfrac{48}{-6} \cdot (-5) =$	**h.** $\dfrac{-12}{16} \cdot \dfrac{-15}{25} \cdot (-3) =$	**i.** $-3 \cdot \dfrac{-8}{-16} \cdot \dfrac{6}{5} \cdot (-5) =$

4. Find the missing factors.

a. $(-8) \cdot$ _____ $\cdot (-2) = -40$	**b.** $2 \cdot$ _____ $\cdot (-4) = 64$	**c.** _____ $\cdot (-3) \cdot 4 = -36$

5. When will the product of many integers be positive? When will it be negative? Experiment! You can, for example, find powers of -1. Find the sign (positive or negative) of the final product, if there are...

    a. three negative factors

    b. four negative factors

    c. five negative factors

    d. six negative factors

6. Based on your work above, fill in:

> When there is an _____ number of factors, the product is negative.
> For example, $(-1)^7$ has seven factors, and thus its value is negative 1.
>
> When there is an _____ number of factors, the product is positive.
> For example, $(-2)^6$ has six factors, and thus its value is (positive) 64.

7. Find the value of the expressions.

a. $(-2)^3 =$	**b.** $(-1)^{18} =$	**c.** $2 \cdot (-1) \cdot (-5) \cdot (-2) \cdot (-3) \cdot (-10) =$
d. $(-1)^{27} =$	**e.** $(-2)^8 =$	**f.** $-3 \cdot 5 \cdot (-1) \cdot (-2) \cdot (-4) \cdot (-2) \cdot (-10) =$

8. Multiply with fractions.

a. $\left(-\dfrac{1}{5}\right)^2 =$	**b.** $\left(-\dfrac{1}{2}\right)^3 =$	**c.** $\left(-\dfrac{2}{3}\right)^4 =$
d. $\left(-\dfrac{1}{2}\right)\left(-\dfrac{7}{8}\right) \cdot \dfrac{4}{3} \cdot \left(-\dfrac{3}{2}\right) =$	**e.** $\left(-\dfrac{1}{2}\right)^5\left(-\dfrac{1}{3}\right)^2 =$	**f.** $\left(\dfrac{2}{3}\right)^2\left(-\dfrac{6}{5}\right) =$

9. Find the value of the expressions.

a. $7 \cdot (-10)^3 \cdot 10^2$	**b.** $(4-6)^3 - (1+2)^3$	**c.** $2(1-5)^2 + 3(-1+2)^4$
d. $\dfrac{(-1)^4}{(-1)^7} + \dfrac{5}{(-2)^3}$	**e.** $\dfrac{-8}{(-2)^4} + \dfrac{(-1)^5}{(-3)^2}$	**f.** $\dfrac{6 \cdot 8}{(-4)^3} + 2 \cdot (-7)^2$

Note on notation. There is a difference between the expressions $(-2)^4$ and -2^4. They do not mean the same.

- In $(-2)^4$, we use brackets to indicate that it is the number -2 that is raised to a power.
  To find its value, we multiply -2 by itself: $(-2)(-2)(-2)(-2) = \mathbf{16}$.

- In contrast, -2^4 means that 2 is raised to the fourth power, and THEN the result of that is negated.
  The exponent only applies to the number next to it. So, we first multiply 2 by itself four times:
  $2 \cdot 2 \cdot 2 \cdot 2 = 16$. Now we apply the negative sign and the final value is $\mathbf{-16}$.

Example 3. Compare the two expressions, and how to find their values.	$2 - (1 - 3^3)$ $= 2 - (1 - 27)$ $= 2 - (-26)$ $= 28$	$2 - (1 - (-3)^3)$ $= 2 - (1 - (-27))$ $= 2 - (28)$ $= -26$

10. Find the value of the expressions.

a. $-5^2 =$ $(-5)^2 =$	**b.** $-10^3 =$ $(-10)^3 =$	**c.** $100 - 2^3 =$ $100 - (-2)^3 =$	**d.** $40 - (30 - 4^3) =$ $40 - (30 - (-4)^3) =$
e. $3 + (-1)^3 =$ $3 - (-1)^3 =$	**f.** $3 - (-2)^2 =$ $3 - (-2^2) =$	**g.** $(3 - 5)^3 =$ $(3^3 - 5^3) =$	**h.** $3 - (-2 - 5)^2 =$ $3 - (-2^2 - 5) =$

11. In each case below, explain why the answer is wrong, and correct it.

    a. $2^5 = 10$ **b.** $-2^4 = 16$ **c.** $(-1)^4 = -1$

12. Find the value of the unknowns.

a. $(-0.2)^x = -0.008$	**b.** $-2^6 = (-4)^y$	**c.** $w^2 = 0.0016$
d. $(-4)^3 = -a^2$	**e.** $x^5 = -\dfrac{1}{32}$	**f.** $x^3 = -\dfrac{8}{27}$

13. Solve, and place the answers to the cross-number puzzle. The negative sign goes in the same box as the first digit of the number.

    Across: **Down:**

    3. $(-10)^3$ 1. $2 \cdot 5^2$

    5. -6^3 2. $-10 \cdot 3^2$

    6. $50 \cdot (-3)^2$ 3. -2^{10}

                       4. $6 \cdot (-3)^3$

                       5. $-8 \cdot (-5)^2$

Laws of Exponents, Part 1

1. **a.** In the expression $2^4 2^3$, both powers have the same base of 2. See if you can find a way to write this expression in a shorter form, as a single power of 2 (using only one exponent).

   Hint: Expand the powers as repeated multiplications..

   b. Do the same with $3^2 3^4$.

   c. Do the same with $a^3 a^9$.

2. Are the following statements true? If not, correct them.

   a. $2^4 2^2 = 2^8$

   b. $2^3 2^3 = 4^6$

   c. $10^3 10^2 = 10^5$

3. Expand the powers by writing out the repeated multiplications. Then simplify. Lastly, write the entire expression as a single power of 4.

   $$\frac{4^7}{4^5} = \underline{\hspace{5cm}} =$$

4. Simplify the expression, writing it as a single power of 5.

   $$\frac{5^5}{5^2} = \underline{\hspace{5cm}} =$$

5. Using the same technique as above, write the expression $\dfrac{x^6}{x^2}$ as a single power of x.

6. Sandra believes that $\dfrac{2^5}{2^4} \cdot 2 = 1$. Is she correct? If not, explain why not.

7. Are the following statements true? Use the table of powers of 3 to help.

   <u>Hint 1:</u> Often estimation is sufficient to see that a statement is wrong.

   <u>Hint 2:</u> To check the veracity of a division statement, you can also use multiplication.

$3^2 = 9$
$3^3 = 27$
$3^4 = 81$
$3^5 = 243$
$3^6 = 729$
$3^7 = 2187$
$3^8 = 6561$

   a. $3^3 + 3^4 = 3^7$ **b.** $3^3 \cdot 3^4 = 3^7$

   c. $\dfrac{3^6}{3^3} = 3^2$ **d.** $\dfrac{3^6}{3^3} = 3^3$

The first law of exponents: product of powers

When we multiply powers <u>with the same base</u>, we are dealing with a repeated multiplication of the same number.

$$5 \cdot 5 \cdot 5 \cdot 5 \;\cdot\; 5 \cdot 5 \cdot 5 \cdot 5 \cdot 5 \cdot 5 \;=\; 5^{10}$$

4 repetitions   6 repetitions   10 repetitions

For example, in $5^4 5^6$, we first multiply five by itself 4 times, and then 6 more times.

We can simply add the exponents: $5^4 5^6 = 5^{10}$. Five is multiplied by itself ten times.

In general, $a^n a^m = a^{n+m}$ for any rational number a, and for any integer exponents n and m.

8. Write each expression as a single power (with one base and one exponent). You don't have to find the value of the expressions.

a. $4^3 4^{11} = $	**b.** $(-7)^{17}\,(-7)^2 = $	**c.** $\left(-\dfrac{2}{3}\right)^2 \left(-\dfrac{2}{3}\right)^5 = $
d. $5 \cdot 5 \cdot 5 \cdot 5^8 = $	**e.** $0.3 \cdot 0.3^2 \cdot 0.3^4 = $	**f.** $a^2 \cdot a \cdot a \cdot a^{12} = $ **g.** $x^m x^7 = $

9. Write each expression as a single power. You don't have to find the value of the expressions.

a. $1000 \cdot 10^4 = $	**b.** $8 \cdot 2^5 = $	**c.** $3^6 \cdot 27 = $	**d.** $4^2 \cdot 64 = $

10. Find the value of x.

a. $6^2\, 6^x\, 6^9 = 6^{21}$	**b.** $b^x b^9 = b^5 b^6$	**c.** $9 \cdot 3^5 = 27 \cdot 3^x$	**d.** $8 \cdot 2^4 = 32 \cdot 2^x$

Example 1. Rewrite $2^9\, 7^4\, 7^2\, 2^5 \cdot 4$ as a product of powers, using as few exponents as possible.

Here, 2 is multiplied by itself nine times, and also five times, and then, 4 is actually $2 \cdot 2$. In total, 2 is multiplied by itself $9 + 5 + 2 = 16$ times.

And 7 is multiplied by itself, first of all four times, and then also two times — a total of 6 times.

In all, the simplified expression is $2^{16}\, 7^6$.

Example 2. Simplify $4x^9 y^4 x^2 y^5 x$.

The variable x is multiplied by itself $9 + 2 + 1 = 12$ times. The variable y is multiplied by itself $4 + 5 = 9$ times.

We cannot do anything about the coefficient 4.

In all, the simplified expression is $4x^{12}\, y^9$.

11. Rewrite each expression as a product of powers, using as few exponents as possible.

a. $10^2\, 8^3\, 10^5\, 8^6$	**b.** $3^3\, 5^6\, 3^3\, 4^3\, 5^2 \cdot 3$	**c.** $24 \cdot 2^3\, 3^6\, 2^5$
d. $25 \cdot 2^4\, 5^2\, 3^2 \cdot 9$	**e.** $64 \cdot 100 \cdot 7 \cdot 2^3$	**f.** $81 \cdot 36 \cdot 3 \cdot 2^3$

12. Simplify.

a. $x^7 y^3 x^3 y^9$	**b.** $2a^3\, a^6\, b^3\, b^3\, ab$	**c.** $5s \cdot r^4 \cdot s \cdot s^3 \cdot s \cdot 6 \cdot r \cdot t^3$

The second law of exponents: quotient of powers. When we divide powers with the same base, such as $\dfrac{7^6}{7^4}$, we can simply **subtract the exponents**: $\dfrac{7^6}{7^4} = 7^{6-4} = 7^2$.

Why is that? When we expand both power expressions, we get this: $\dfrac{7^6}{7^4} = \dfrac{7 \cdot 7 \cdot 7 \cdot 7 \cdot 7 \cdot 7}{7 \cdot 7 \cdot 7 \cdot 7}$.

Now we can cancel out many sevens, and we are left with: $\dfrac{\cancel{7} \cdot \cancel{7} \cdot \cancel{7} \cdot \cancel{7} \cdot 7 \cdot 7}{\cancel{7} \cdot \cancel{7} \cdot \cancel{7} \cdot \cancel{7}} = 7^2$.

In general, $\dfrac{a^m}{a^n} = a^{m-n}$ for any nonzero a and any exponents m and n.

13. Write out the multiplications in the numerator and in the denominator. Then simplify.

$$\frac{a^6}{a^3} = \frac{\square \cdot \square \cdot \square \cdot \square \cdot \square \cdot \square}{\square \cdot \square \cdot \square} =$$

14. Write each expression as a single power (with one base and one exponent).

a. $\dfrac{8^9}{8^3}$	**b.** $\dfrac{(-7)^{12}}{(-7)^5}$	**c.** $\dfrac{3 \cdot 3^6}{3^4}$	**d.** $\dfrac{0.4 \cdot 0.4^{10}}{0.4^5 \cdot 0.4}$
e. $\dfrac{y^{25}}{y^{14}}$	**f.** $\dfrac{7x^4 x^6}{x^9}$	**g.** $\dfrac{16 \cdot 4^5}{4^2 \cdot 64}$	**h.** $\dfrac{125 \cdot 5^6}{5^2 \cdot 25}$

15. Find the value of the unknowns.

a. $\dfrac{11^{10}}{11^x} = 11^7$	**b.** $\dfrac{5.4^y}{5.4^5} = 5.4^4$	**c.** $\dfrac{(-2)^w}{(-2)^7} = -32$	**d.** $\dfrac{(-6)^x}{(-6)^3} = 36$

16. Andrew believes that the answer to this is 3. Explain why his thinking is wrong.

> How many times greater is 2^6 than 2^2?

17. How can we simplify, if we have powers with two different bases in a fractional expression? Simplify the expressions. You can expand the power expressions to help you.

a. $\dfrac{8^6 \cdot 5^3}{8^3 \cdot 5^2}$ 　　　　　　**b.** $\dfrac{x^6 y^8}{y^3 x^3}$ 　　　　　　**c.** $\dfrac{3x^4 y^6}{5xy^2}$

18. Find the expressions that have the value 2^6.

a. $\dfrac{2^{12}}{2^2}$	**b.** 12	**c.** $\dfrac{2^9}{2^3}$	**d.** $2^2 \cdot 2^3$	**e.** 6^2	**f.** $\dfrac{2^4}{1/4}$

Zero and Negative Exponents

What could be meant by a number raised to the zeroth power, such as 5^0 or 12^0? Thinking of repeated multiplication doesn't work. Some people suggest they have a value of zero, but let's look at some patterns. Fill in each pattern, dividing by the base number in each step as you go down.

2^3	8		5^3	125		10^3	1000		a^3	$a \cdot a \cdot a$
2^2		$\div 2$	5^2		$\div 5$	10^2		$\div 10$	a^2	$a \cdot a$
2^1		$\div 2$	5^1		$\div 5$	10^1		$\div 10$	a^1	
2^0		$\div 2$	5^0		$\div 5$	10^0		$\div 10$	a^0	

The natural, though surprising, result is also the definition: $a^0 = 1$ for all numbers a.

You will see more justification for this definition in the exercises of this lesson.

(Mathematicians either leave 0^0 undefined, or define it as 1, depending on the context.)

1. The product law of powers states that $a^m a^n = a^{m+n}$. Let's apply that law when $n = 0$.

   We get: $a^m \cdot a^0 = a^{m+0} = a^m$. From that we essentially have: $a^m \cdot a^0 = a^m$.

   So, if we wish the product rule to work with a zero exponent, what must be the value of a^0?

2. Apply the quotient law of powers, $\dfrac{a^m}{a^n} = a^{m-n}$, when $m = 4$ and $n = 4$:

   How does that show you that the value of a^0 must be 1?

3. **a.** Apply the quotient law of powers, $\dfrac{a^m}{a^n} = a^{m-n}$, $a = 3$, $m = 2$, and $n = 3$:

   What does that suggest for the value of 3^{-1}?

   b. Try it also when $a = 5$. What is the suggested value of 5^{-1}?

4. What could make sense for a definition of a negative exponent?

   Let's try patterns once again.

   Continue the patterns, to see what happens.

2^3			5^3			10^3		
2^2		$\div 2$	5^2		$\div 5$	10^2		$\div 10$
2^1		$\div 2$	5^1		$\div 5$	10^1		$\div 10$
2^0		$\div 2$	5^0		$\div 5$	10^0		$\div 10$
2^{-1}		$\div 2$	5^{-1}		$\div 5$	10^{-1}		$\div 10$
2^{-2}		$\div 2$	5^{-2}		$\div 5$	10^{-2}		$\div 10$
2^{-3}		$\div 2$	5^{-3}		$\div 5$	10^{-3}		$\div 10$

You saw on the previous page how the values for powers with negative exponents become fractions if we follow patterns. Indeed that is what the definition is based on. The power a^{-n} is defined as the reciprocal of a^n:

$$a^{-n} = \frac{1}{a^n}$$, where a is any number except zero, and n is an integer.

a^2	$a \cdot a$
a^1	a
a^0	1
a^{-1}	$1/a$
a^{-2}	$1/a^2$
a^{-3}	$1/a^3$

$\div a$ $\div a$ $\div a$ $\div a$ $\div a$

Example 1. $3^{-4} = \dfrac{1}{3^4} = \dfrac{1}{81}$, and $12^{-2} = \dfrac{1}{12^2} = \dfrac{1}{144}$.

Example 2. Simplify $3 \cdot 7^{-1}$. Note that the exponent only applies to 7, not to 3.

First of all, $7^{-1} = \dfrac{1}{7^1} = \dfrac{1}{7}$. The original expression $3 \cdot 7^{-1}$ then equals $\dfrac{3}{7}$.

5. Simplify, giving your final answer without any exponents (as a fraction or a whole number).

a. $6^{-2} =$	**b.** $2^{-4} =$	**c.** $5^{-3} =$	**d.** $7 \cdot 10^{-1} =$	**e.** $10 \cdot 3^{-3} =$

6. Write each fraction in the form $a \cdot b^{-n}$, where a, b, and n are whole numbers, and b is as small as possible.

a. $\dfrac{2}{5} =$ **b.** $\dfrac{5}{8} =$ **c.** $\dfrac{11}{9} =$ **d.** $\dfrac{37}{1000} =$

7. Does the definition of a negative exponent work when $n = 0$?

$$a^{-n} = \frac{1}{a^n}$$

Example 3. The law about the product of powers states that $a^n a^m = a^{n+m}$. This law works with negative exponents, too. For example, $8^{-6} 8^4 = 8^{-6+4} = 8^{-2}$.

Note how this works when expanded: $8^{-6} 8^4 = \dfrac{8^4}{8^6} = \dfrac{8 \cdot 8 \cdot 8 \cdot 8}{8 \cdot 8 \cdot 8 \cdot 8 \cdot 8 \cdot 8} = \dfrac{1}{8^2} = 8^{-2}$.

8. Simplify. Give your answers without negative exponents.

a. $5 \cdot 2^8 2^{-5} =$	**b.** $5 \cdot 2^{-2} 2^{-3} \cdot 3 =$	**c.** $x^{-3} x^2 =$
d. $y \cdot y^{-2} y^5 =$	**e.** $3a^{-4} a^7 a^{-5} =$	**f.** $2b^5 a^{-3} a^{-2} =$

9. Find the value of the unknowns.

a. $5^x = \dfrac{1}{125}$	**b.** $10^x = \dfrac{1}{1000000}$	**c.** $19^{-5} = \dfrac{19^2}{19^x}$	**d.** $3^2 3^b = \dfrac{1}{27}$
e. $6^{11} 6^z = 6^7$	**f.** $21^7 21^a = 1$	**g.** $8^7 8^y = 8^{-3} 8^{10}$	**h.** $9^2 3^5 = 3^{-3} 9^x$

> **Example 4.** Simplify $(-4)^{-2}$.
>
> Here we need to be careful with the negatives. We have -4 raised to the power of -2. However, the negative exponent does *not* signify a negative number; it signifies that we need to use the *reciprocal of the base* raised to the corresponding positive exponent. Notice carefully: $(-4)^{-2} = \dfrac{1}{(-4)^2} = \dfrac{1}{16}$.

10. How is $(-4)^{-2}$ different from -4^{-2}?

11. Find the values of: **a.** 2^{-3} **b.** -2^{-3} **c.** $(-2)^{-3}$ **d.** $(-2)^3$

12. Simplify, giving your final answer without any exponents (as a fraction or a whole number).

a. $(-5)^{-2} =$	**b.** $(-3)^{-3} =$	**c.** $(-2)^{-5}(-2)^2 =$
d. $53 \cdot (-100)^{-1} =$	**e.** $7 \cdot (-4)^{-1}(-4)^{-2} =$	**f.** $(-10)^3(-10)^{-2} \cdot 5 =$

13. Find the value of the unknowns.

a. $(-3)^x = \dfrac{1}{81}$	**b.** $(-10)^y = -\dfrac{1}{100000}$	**c.** $(-2)^4(-2)^w = -\dfrac{1}{8}$

14. Which is more? Note: you don't need to calculate the actual values of the powers.

a. 6^2 or 2^{-6}	**b.** -7^2 or 2^{-7}	**c.** 9^{-8} or 9^{-7}	**d.** $(-4)^{-8}$ or $(-4)^{-7}$

15. Simplify, giving your answer without any exponents.

a. $10^{-1}5^2 =$	**b.** $3^{-1}2^{-2} =$	**c.** $5^{-8} \cdot 12 \cdot 5^8 =$
d. $10 \cdot 3^2 5^{-1} 4^{-2} =$	**e.** $6 \cdot 2^{-3} 5^{-1} =$	

16. Find the expressions that have a value of 1.

a. $3^2 \cdot (-2)^{-3}$	**b.** $4^2 \cdot 2^{-4}$	**c.** $11^{-6}11^6$
d. $\dfrac{2^2 2^5}{2^4 2^3}$	**e.** $2^{-4} \cdot \dfrac{2^4}{64}$	**f.** $5^{-1} \cdot \dfrac{5^3}{25}$

More on Negative Exponents

The expression 5^{-2} is a fraction (which?). If that expression is in the *denominator*, we will get **a complex fraction** (a fraction that contains a fraction in the numerator, denominator, or both).

For example, $\dfrac{3}{5^{-2}} = \dfrac{3}{\dfrac{1}{5^2}} = \dfrac{3}{\dfrac{1}{25}}$. To simplify this expression further, we will use fraction division.

Recall that that involves a multiplication by the reciprocal of the divisor: $\dfrac{3}{\dfrac{1}{25}} = 3 \div \dfrac{1}{25} = 3 \cdot 25 = 75$.

1. Find the value of the expressions. Be on the lookout for a shortcut.

a. $\dfrac{1}{9^{-2}} =$	b. $\dfrac{1}{3^{-3}} =$	c. $\dfrac{5}{2^{-4}} =$
d. $\dfrac{1}{x^{-5}} =$	e. $\dfrac{7}{a^{-4}} =$	f. $\dfrac{1}{x^{-n}} =$

Example 1. Simplify $\dfrac{3^{-2}}{4 \cdot 3^{-5}}$.

The rule for subtracting exponents in a quotient works even when the exponents are negative.

We get $\dfrac{3^{-2}}{4 \cdot 3^{-5}} = \dfrac{1}{4} \cdot 3^{-2 - (-5)} = \dfrac{1}{4} \cdot 3^3 = \dfrac{27}{4} = 6\dfrac{3}{4}$.

Another way is to use the shortcuts that $\dfrac{1}{3^{-5}} = 3^5$ and $3^{-2} = \dfrac{1}{3^2}$. This means that $\dfrac{3^{-2}}{3^{-5}} = \dfrac{3^5}{3^2}$.

It's like you flip the exponents from negative to positive by moving the powers from numerator to denominator or vice versa! So then $\dfrac{3^{-2}}{4 \cdot 3^{-5}} = \dfrac{1}{4} \cdot \dfrac{3^5}{3^2} = \dfrac{3^3}{4} = \dfrac{27}{4} = 6\dfrac{3}{4}$.

2. Simplify, writing the expressions with the least amount of exponents, and without negative exponents.

a. $\dfrac{7^{-3}}{7^{-6}} =$ b. $\dfrac{2x^5}{x^{-1}} =$ c. $\dfrac{3y^{-8}}{6y^{-4}} =$ d. $\dfrac{24s^{-2}}{6w^{-7}} =$

3. Find the value of the expressions.

a. $\dfrac{9 \cdot 2^{-4}}{5 \cdot 2^{-2}} =$	b. $\dfrac{3 \cdot 10^{-1}}{5 \cdot 10^{-3}} =$	c. $\dfrac{10 \cdot 5^{-1}}{3 \cdot 5^2} =$

Example 2. Rewrite the expression $\dfrac{3 \cdot 10^{-2} \cdot 3^5}{6 \cdot 10^3 \cdot 3^{-2}}$ with the fewest bases and exponents possible.

Note, first of all, that this fractional expression is equal to $\dfrac{3}{6} \cdot \dfrac{10^{-2}}{10^3} \cdot \dfrac{3^5}{3^{-2}}$.

Simplifying the powers of 10, we get $\dfrac{10^{-2}}{10^3} = 10^{-2-3} = 10^{-5}$.

Simplifying the powers of 3, we get $\dfrac{3^5}{3^{-2}} = 3^{5-(-2)} = 3^7$.

The original expression now becomes $\dfrac{3}{6} \cdot \dfrac{10^{-2}}{10^3} \cdot \dfrac{3^5}{3^{-2}} = 0.5 \cdot 10^{-5} \cdot 3^7$.

If preferred, this can be written with the fraction line instead, as: $\dfrac{3^7}{2 \cdot 10^5}$.

4. Rewrite each expression using powers, minimising the number of exponents and of bases, and without negative exponents (use the fraction line).

a. $\dfrac{2^{-5} \cdot 7^2}{2^{-2} \cdot 7^{-3}} =$

b. $\dfrac{2 \cdot 5^3 \cdot 3^{-7}}{5^{-3} \cdot 3^4} =$

c. $\dfrac{2 \cdot 3^{-5} \cdot 10^2}{7 \cdot 10^{-3} \cdot 3^{-6}} =$

d. $\dfrac{7 \cdot 7^{-5} \cdot 2^{-3}}{4 \cdot 7^3 \cdot 2^3} =$

e. $\dfrac{x^6 y^2}{x^3 y^3}$

f. $\dfrac{2a^4 b^7}{b^3 a^{-2}} =$

g. $\dfrac{2a^2 b^{-2}}{3a^{-3} b^3} =$

h. $\dfrac{5x^{-4} y^2}{15 y^3 x^{-6}} =$

5. Use the table of powers of 4 to determine the value of the expressions.

a. $4^5 \cdot 4^4$

b. $(-4)^3 \cdot (-4)^2 \cdot (-4)^2$

c. $\dfrac{4^3 \, 4^6}{4^4}$

d. $\dfrac{4^3 \, 4^{-6}}{4^{-2} \, 4^4}$

e. $4^2 \cdot 4^{12} \cdot 4^{-6}$

f. $-4^2 \cdot 4^{-3} \cdot 4^{-2}$

g. $\dfrac{4 \cdot 4^{-4}}{4^{-3} \, 4^2}$

h. $\dfrac{4^{-3} \, 4^2}{4^{-5} \, 4^4}$

4^3	$= 64$
4^4	$= 256$
4^5	$= 1024$
4^6	$= 4096$
4^7	$= 16\,384$
4^8	$= 65\,536$
4^9	$= 262\,144$
4^{10}	$= 1\,048\,576$

Laws of Exponents, Part 2

1. Use the definition for a positive exponent (repeated multiplication) and for a negative exponent (reciprocal of the same power with a positive exponent), and see if you can write each expression using a single exponent.

a. $(9^5)^2 = $ ☐ $\cdot$ ☐ $= $

b. $(a^4)^3 = $ ☐ $\cdot$ ☐ $\cdot$ ☐ $= $

c. $(x^{-1})^3 = \left(\dfrac{1}{☐}\right)^☐ = \dfrac{1}{☐} \cdot \dfrac{1}{☐} \cdot \dfrac{1}{☐} = $

What you probably found in exercise 1 suggests a third law of exponents — or really just a shortcut:
When a power is raised to a power, you can multiply the exponents.

The third law of exponents: $(a^n)^m = a^{nm}$ for any number a, and for any integer exponents n and m.

Example 1. The expression $(9^3)^2$ means that we multiply 9^3 by itself two times: $9^3 \cdot 9^3$.
Expanding those powers, we get $(9 \cdot 9 \cdot 9) \cdot (9 \cdot 9 \cdot 9) = 9^6$. So, the shortcut $(9^3)^2 = 9^6$ makes sense.

Example 2. Is $(x^5)^{-2}$ really equal to x^{-10}? Let's check this out.

To simplify $(x^5)^{-2}$, we first use the definition of a negative exponent: $(x^5)^{-2} = \dfrac{1}{(x^5)^2}$.

Now we expand $(x^5)^2$ and get $\dfrac{1}{(x^5)^2} = \dfrac{1}{x^5 x^5} = \dfrac{1}{x^{10}}$. This, in turn, is equal to x^{-10}. So, yes, $(x^5)^{-2} = x^{-10}$.

---

2. Write each expression using a single exponent.

a. $(8^2)^4 =$	**d.** $x^6 x^{-3} =$	**g.** $((-2)^{-3})^5 =$	**j.** $s^{-1} s^{-7} =$
b. $8^2 8^4 =$	**e.** $(x^6)^{-3} =$	**h.** $(-2)^{-3}(-2)^5 =$	**k.** $(s^{-1})^{-7} =$
c. $\dfrac{8^2}{8^4} =$	**f.** $\dfrac{x^6}{x^{-3}} =$	**i.** $\dfrac{(-2)^{-3}}{(-2)^5} =$	**l.** $\dfrac{s^{-1}}{s^{-7}} =$

3. Find the value of each unknown.

a. $1\,000\,000 = (10^x)^3$	**b.** $256 = (2^2)^x$	**c.** $(5^x)^3 = \dfrac{1}{5^{15}}$	**d.** $(2^x)^2 = \dfrac{1}{64}$

4. Change the base of each power expression, following the example.

a.	**b.**	**c.**	**d.**
$27^6 = (3^3)^6 = 3^{18}$	$100^6 = (10^☐)^6 = 10^☐$	$125^3 = (5^☐)^☐ = 5^☐$	$4^{12} = (2^☐)^☐ = 2^☐$

With brackets:	Meaning	Without brackets:	Meaning
$(-3)^2$	Negative three squared. Value: 9.	-3^2	The opposite of three squared. Value: -9.
$(9^3)^2$	9^3 multiplied by itself: $(9^3)(9^3) = 9^6$.	9^{3^2}	The exponent 2 applies *only* to the base of 3, not to nine. To simplify it, we first solve $3^2 = 9$. The original expression becomes $9^{3^2} = 9^9$.
$(4b)^3$	The term $4b$ is cubed.	$4b^3$	Only b is cubed, 4 is not.

Note: The exponent *only* applies to the number or variable (the base) right next to it. We need to use brackets to indicate otherwise.

5. In each case, find the value of the two similar expressions. Think carefully what is different about them.

a.	**b.**	**c.**	**d.**
$(-7)^2 =$	$2 \cdot 5^3 =$	$3 \cdot 4^{-2} =$	$\left(\dfrac{2}{3}\right)^3 =$
$-7^2 =$	$(2 \cdot 5)^3 =$	$(3 \cdot 4)^{-2} =$	$\dfrac{2^3}{3} =$

6. Find the values of the expressions and place them in the cross-number puzzle. The negative sign goes in the same box as the first digit of the number.

Across:

1. $540 \cdot 2^{-2}$

3. $-2 \cdot 12^2$

5. $(-2)^{-3}(-2)^9$

8. $3 \cdot 3^6 \cdot 3^{-2}$

Down:

1. $2 \cdot (2^2)^3$

2. $2 \cdot 3^3$

3. $2^3 \cdot (-3)^3$

4. 3^{2^2}

6. $126 \cdot 3^{-1}$

7. $-104 \cdot (2^{-1})^3$

7. **a.** If $a = 2$ and $b = 9$, what is the value of $(a + b)^2$? Of $a^2 + b^2$?

b. Use the visual model to explain why $(a + b)^2$ does not equal $a^2 + b^2$.

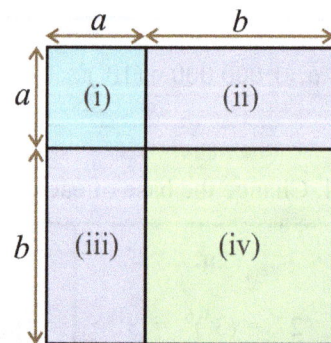

Example 3. What is the difference between the expressions $5x^{-2}$ and $(5x)^{-2}$?

In the expression $5x^{-2}$, the number five is *not* raised to any power. This expression means $5 \cdot x^{-2}$.

We *can* write it with a positive exponent: $5x^{-2} = 5 \cdot \dfrac{1}{x^2} = \dfrac{5}{x^2}$.

In the expression $(5x)^{-2}$, the entire term $5x$ is raised to the power of -2. We can write it

using a positive exponent this way: $(5x)^{-2} = \dfrac{1}{(5x)^2} = \dfrac{1}{(5x) \cdot (5x)} = \dfrac{1}{25x^2}$.

8. Use the definition of an exponent as repeated multiplication to write an equivalent expression to $(xy)^3$.

$(xy)^3 = $ ▢ $\cdot$ ▢ $\cdot$ ▢ $= $ ▢ $\cdot$ ▢

9. Write an equivalent form of each expression without brackets and without negative exponents.

a. $(4x)^2$	**b.** $(2a)^3$	**c.** $2(x^2)^3$
d. $(ab)^4$	**e.** $(3y)^{-2}$	**f.** $(2x)^{-3}$

10. Robert and Xavier worked on simplifying $(5a^3)^2$. Robert got $5a^5$ and Xavier got $10a^6$.
    Who is correct? Explain how you know.

11. What is the difference between expressions $(5 \text{ m})^3$ and 5 m^3?
    (Here, "m" refers to metres.)
    Hint: consider the illustration on the right.

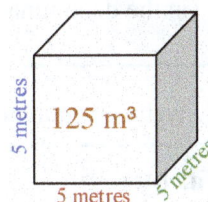
5 metres
125 m³
5 metres
5 metres

12. Join equivalent expressions with a line.
    Some expressions will not be joined.

$\dfrac{4x^2}{y}$	$7x^6$	$4x^2$	$-4x^2$
$(-2x)^2$	$\dfrac{(2x)^2}{y}$	$\left(\dfrac{x}{y}\right)^2$	$7x^9$
$\dfrac{x^2}{y^2}$	$\dfrac{x^2}{y}$	$7x^{3^2}$	$\dfrac{2x^2}{y}$

Laws of Exponents, Part 3

Summary of exponent laws

Here is a list of definitions and common laws, or rules, of exponents (really, they're shortcuts that easily follow from the definitions). We have studied most of them in the previous lessons.

Definitions		Rules (shortcuts)	
Exponent:	m times $a^m = a \cdot a \cdot a \cdot ... \cdot a \cdot a$	Product of powers:	$a^m \cdot a^n = a^{m+n}$
Exponent of one:	$a^1 = a$	Quotient of powers:	$\dfrac{a^m}{a^n} = a^{m-n}$
Zero exponent:	$a^0 = 1$	Power of a power:	$(a^n)^m = a^{nm}$
Negative exponent:	$a^{-n} = \dfrac{1}{a^n}$	Power of a product:	$(ab)^m = a^m b^m$
		Power of a quotient:	$\left(\dfrac{a}{b}\right)^m = \dfrac{a^m}{b^m}$

Note: m and n are integers, and a and b are nonzero numbers.

Example 1. To simplify $(2x^3)^4$, we first use the power of a product rule. In other words, we apply the outside exponent (4) to both 2 and x^3.

We get $(2x^3)^4 = 2^4 \cdot (x^3)^4$. Now we simplify 2^4 and also use the power of a power rule: $2^4 \cdot (x^3)^4 = \mathbf{16x^{12}}$.

Example 2. To simplify $(-b^3)^5$, think of it as $(-1 \cdot b^3)^5$. Now we apply the outside exponent of 5 to both -1 and b^3. We get $(-b^3)^5 = (-1)^5 \cdot (b^3)^5 = -1 \cdot b^{15} = \mathbf{-b^{15}}$.

Example 3. To simplify $\left(\dfrac{3x}{y^2}\right)^2$, we first use the shortcut for the power of a quotient: $\left(\dfrac{3x}{y^2}\right)^2 = \dfrac{(3x)^2}{(y^2)^2}$

Now, we need to use the power of a product rule (in the numerator) and the power of a power rule

(in the denominator). We get $\dfrac{(3x)^2}{(y^2)^2} = \dfrac{3^2 x^2}{y^4} = \dfrac{9x^2}{y^4}$.

1. Write an equivalent expression using the exponent rules, and without negative exponents.

a. $(4y)^3 =$	**g.** $\left(\dfrac{x}{y^2}\right)^3 =$
b. $(11x)^2 =$	**h.** $\left(\dfrac{7g}{8h}\right)^2 =$
c. $(3ab)^3 =$	
d. $(-2w)^3 =$	**i.** $\left(\dfrac{ab}{3c^5}\right)^4 =$
e. $(-x^2)^5 =$	**j.** $\left(\dfrac{-2b}{b^2}\right)^5 =$
f. $(2x)^4 \cdot (-3x)^2 =$	**k.** $\left(\dfrac{-x^2}{5x}\right)^3 =$

2. Let's practise some with negative exponents, also. Simplify, and write an equivalent expression that does not have any negative exponents.

a. $(5x^2)^{-1} =$	**e.** $(10x^{-3})^3 =$	**i.** $(7e^2d^{-4})^2 =$
b. $(4y^3)^{-2} =$	**f.** $(ab^2)^{-2} =$	**j.** $(-2x)^3 \cdot (3x)^{-2} =$
c. $(3a^4)^{-3} =$	**g.** $(5x^2y)^{-3} =$	**k.** $(-3y^2)^{-4} \cdot y^5 =$
d. $(7b^{-4})^2 =$	**h.** $(-3w^{-2})^5 =$	**l.** $(-3c)^3 \cdot (5c)^{-2} =$

3. One of these simplification processes has an error. Which one? Correct it.

a. $\left(\dfrac{-4x}{-y^3}\right)^2 = \dfrac{(-4x)^2}{(-y^3)^2} = \dfrac{16x^2}{y^6}$.	**b.** $\left(\dfrac{4ab}{-3b^3}\right)^3 = \dfrac{12a^3b^3}{-27b^6} = -\dfrac{4a^3}{9b^3}$.

4. A challenge! Find the value of each unknown.

a. $(4^x)^3 = 16^6$	**b.** $81^4 = (3^x)^2$	**c.** $(32 \cdot 5^x)^{-3} = \dfrac{1}{10^{15}}$

5. Join equivalent expressions with a line. Some expressions will not be joined.

$\left(\dfrac{6x}{y}\right)^2$	$8a^3b^{-6}$	$\dfrac{6x^2}{y^2}$	$36x^2y^2$
$36x^2y^{-2}$	$\dfrac{(6x)^2}{y^{-2}}$	$\left(\dfrac{2a}{b^2}\right)^3$	$8a^6b^6$
$\dfrac{6a^3}{b^5}$	$\dfrac{2a^2}{b^6}$	$-6x^2y^2$	$\dfrac{8a^3}{b^6}$

Puzzle Corner

What could a *fractional* exponent mean?

We can use the law of product of powers to try to find the answer.

For example: $9^{\frac{1}{2}} \cdot 9^{\frac{1}{2}} = 9^{\frac{1}{2} + \frac{1}{2}} = 9^1 = 9$

a. Based on this equation, what is the value of $9^{\frac{1}{2}}$?

b. In a similar manner, find the value of $100^{\frac{1}{2}}$ and $64^{\frac{1}{2}}$.

Scientific Notation: Large Numbers

The mass of the planet earth is about 5 972 370 000 000 000 000 000 000 kilograms.
The mass of a single proton (a particle inside an atom) is about 0.0000000000000000000000000016726 kg.

When a number is either extremely large or extremely small, it is hard to keep track of its zeros and to get a grasp of exactly how big or how small it is. To help these issues, we can use **scientific notation**.

In scientific notation, a number is written as a product of a decimal number and a power of ten. For example, the mass of the earth is written as $5.97237 \cdot 10^{24}$ kg, and the mass of a proton as $1.6726 \cdot 10^{-27}$ kg.

A positive number written in scientific notation is in the form of $a \cdot 10^n$, where n is an integer, and a is a decimal number so that $1 \leq a < 10$. Study the examples:

Scientific Notation	equals	Decimal notation
$7 \cdot 10^5$	$7 \cdot 100\ 000$	700 000
$2.83 \cdot 10^6$	$2.83 \cdot 1\ 000\ 000$	2 830 000
$1.032 \cdot 10^8$	$1.032 \cdot 100\ 000\ 000$	103 200 000

Note:
- The power of ten gives you the largest place value for the number. For example, in $2.83 \cdot 10^6$, the power of ten is 10^6, and it tells us that the largest place is the millions place.

- For positive exponents (large numbers), the number of digits in each number is just one more than what the exponent indicates. For example, $2.83 \cdot 10^6 = 2\ 830\ 000$ has *seven* digits in total.

- The factor a has to be at least 1 but less than 10. It is for that reason that $0.95 \cdot 10^5$ and $237 \cdot 10^3$ are not written correctly in scientific notation.

Example 1. In $5.078 \cdot 10^{11}$, the power of ten is 10^{11}, and thus the largest place is the hundred billions place. The number will have a total of 12 digits (11 plus 1). The <u>digit 5 goes to the hundred billions place</u>, and the other digits (0, 7, and 8) follow it. In decimal notation, $5.078 \cdot 10^{11}$ is written as 507 800 000 000.

1. Write the numbers in decimal notation.

Scientific Notation	Decimal notation	Scientific Notation	Decimal notation
$6 \cdot 10^5$		$8.904 \cdot 10^3$	
$2.5 \cdot 10^5$		$1.5594 \cdot 10^8$	
$2.03 \cdot 10^6$		$3.6002 \cdot 10^{11}$	

2. **a.** Eric said that $3.58 \cdot 10^9$ has nine zeros, like this: 358 000 000 000. Is he correct? If not, write the number correctly in decimal notation.

   b. For what single-digit values of a will the number $a \cdot 10^9$ have exactly nine zeros?

Example 2. Write 25 600 in scientific notation.

The largest place value is <u>ten thousands,</u> so the power of ten to use is 10^4. Next, write the digits from 25 600, excluding the trailing zeros, and put a decimal point after the first digit. We get 2.56. So $25\ 600 = \mathbf{2.56 \cdot 10^4}$.

Example 3. Write 6 078 500 000 in scientific notation.

The largest place value is billions or 10^9. When we exclude the trailing zeros, the digits are 60785. We put a decimal point right after 6 to get 6.0785. So 6 078 500 000 is $\mathbf{6.0785 \cdot 10^9}$.

3. Write the numbers in scientific notation.

   a. 13 000   **b.** 204 000   **c.** 4 506 000

   d. 45 280 000   **e.** 9 700 500 000   **f.** 405 100 000 000

4. Compare the numbers, writing < or > in the box.

   a. $9 \cdot 10^7$ ☐ $2 \cdot 10^8$ **b.** $3.4 \cdot 10^7$ ☐ $9.5 \cdot 10^6$ **c.** $7.82 \cdot 10^7$ ☐ $2.87 \cdot 10^8$

   d. $2 \cdot 10^5$ ☐ 32 000   **e.** 405 000 ☐ $4.5 \cdot 10^5$ **f.** 2 190 000 ☐ $1.19 \cdot 10^7$

Example 4. $0.86 \cdot 10^4$ is *not* written correctly in scientific notation, because the factor 0.86 should be more than 1, yet less than 10. We need to use 8.6 instead of 0.86. Since changing 0.86 to 8.6 makes it *ten* times larger, then 10^4 must become ten times smaller, which means it becomes 10^3. So, $0.86 \cdot 10^4 = 8.6 \cdot 10^3$.

5. To write $210 \cdot 10^5$ correctly in scientific notation, Charlie wrote it as $2.1 \cdot 10^7$ and Emily wrote it as $2.1 \cdot 10^3$. Who is correct? Explain why.

6. **a.** Match the expressions and numbers with the same value.

   b. Circle the ones that are written in scientific notation correctly.

$60 \cdot 10^8$	60 000 000	$0.6 \cdot 10^5$
$6 \cdot 10^4$	$600 \cdot 10^7$	$6 \cdot 10^9$
6 000 000 000	$6 \cdot 10^8$	60 000

7. Rewrite in scientific notation correctly.

   a. $26 \cdot 10^6$ **b.** $0.9 \cdot 10^5$ **c.** $358 \cdot 10^4$

   d. $0.208 \cdot 10^7$ **e.** $0.02 \cdot 10^8$ **f.** $10.1 \cdot 10^6$

8. Compare the numbers, writing <, >, or = in the box.

   a. $5 \cdot 10^7$ ☐ $50 \cdot 10^6$ **b.** 350 000 ☐ $3.5 \cdot 10^6$ **c.** $0.8 \cdot 10^5$ ☐ $8 \cdot 10^6$

9. An imaginary country has 20 million citizens, of which 80% are adults. The country is considering a universal income program where each adult citizen would get a basic income of $20 000 per year.

   a. Find the cost of this program for one year. Use scientific notation in your calculation. (Do not use a calculator.)

   b. How should you give your final answer? What number format / unit makes most sense?

10. Let's scale it up! How much would this program cost if it involved the entire population of the planet? Use 8 billion as a rough estimate for the world population, and an estimate that 25% of those are children, not included in the program. (Do not use a calculator.)

11. Complete the "cross-number puzzle" by writing the numbers in decimal notation.

    Across:

    a. $8.74 \cdot 10^5$

    d. $3 \cdot 10^2$

    f. $5.032 \cdot 10^5$

    g. $2.99 \cdot 10^7$

    Down:

    a. $8.93 \cdot 10^3$

    b. $4.205 \cdot 10^4$

    c. $7.4 \cdot 10^6$

    e. $5.2 \cdot 10^4$

Scientific Notation: Small Numbers

When writing very small numbers in scientific notation, the exponent is negative. Recall that negative exponent does not signify a negative number, but a *fraction*: a reciprocal of the corresponding power of ten with a positive exponent. For example, $10^{-4} = \frac{1}{10^4} = \frac{1}{10000}$. As a decimal, this is 0.0001.

Example 1. To write 0.00034 in scientific notation, we need to use 3.4 as the decimal that is multiplied by a power of ten. (Why?) Now note in which place the digit 3 is: it is in ten-thousandths place, which is the *fourth* digit after the decimal point. This means we use $\frac{1}{10000} = \frac{1}{10^4} = 10^{-4}$ as the power of ten.

So, $0.00034 = 3.4 \cdot 10^{-4}$.

Example 2. To write $7.64 \cdot 10^{-6}$ in decimal notation, we note that the digit 7 has to be in the place indicated by the power of ten, which means 7 will be in the millionths place. The other digits will follow. The millionths place is the sixth decimal digit after the decimal point. So, $7.64 \cdot 10^{-6} = 0.00000764$.

1. Write the numbers given in scientific notation in decimal notation, and vice versa.

Scientific Notation	Decimal notation	Scientific Notation	Decimal notation
$3 \cdot 10^{-5}$			0.0000002388
	0.0008	$8.2 \cdot 10^{-4}$	
	0.00000203		0.00000000308
$6.108 \cdot 10^{-8}$		$4.539 \cdot 10^{-7}$	

2. Eric said that $7.61 \cdot 10^{-9}$ has a total of nine zeros, like this: 0.00000000761. Is he correct?

3. Compare the numbers, writing < or > in the box. How can you tell which of them is greater, without writing them in decimal notation?

a. $2 \cdot 10^{-7}$ ☐ $5 \cdot 10^{-8}$

b. $3 \cdot 10^{-9}$ ☐ $3 \cdot 10^{-7}$

c. $7.82 \cdot 10^{-5}$ ☐ 0.000075

d. $4 \cdot 10^{-4}$ ☐ 0.00046

e. $7 \cdot 10^{-4}$ ☐ 0.0065

f. 0.00000078 ☐ $2.8 \cdot 10^{-8}$

4. Write in order from smallest to greatest: $5.6 \cdot 10^7$ 10^{-6} 0.0003 10^8 $6 \cdot 10^7$ 0.00002 $9 \cdot 10^{-7}$

5. A 70-kg male body contains approximately 7 000 000 000 000 000 000 000 000 000 atoms. Of these, approximately $3.9 \cdot 10^{25}$ are nitrogen atoms, $1.61 \cdot 10^{27}$ are oxygen atoms, $8.03 \cdot 10^{26}$ are carbon atoms, and $4.22 \cdot 10^{27}$ are hydrogen atoms.

   a. Write the number of atoms in the male 70-kg body using scientific notation.

   b. Are there more oxygen or carbon atoms in a human body?

> **Example 3.** To write $77 \cdot 10^{-4}$ correctly in scientific notation, we need to use 7.7 instead of 77. Since changing 77 to 7.7 makes it *ten* times smaller, then 10^{-4} must become ten times larger, which means we multiply it by 10 and it becomes 10^{-3}. So, $77 \cdot 10^{-4} = 7.7 \cdot 10^{-3}$.

6. One gold atom weighs about $326.964 \cdot 10^{-24}$ grams. Write this number in scientific notation correctly.

7. Rewrite the numbers in scientific notation correctly.

   a. $89 \cdot 10^{-5}$ **b.** $479 \cdot 10^{-6}$

   c. $0.3 \cdot 10^{-4}$ **d.** $208 \cdot 10^{-9}$

   e. $0.045 \cdot 10^{-8}$ **f.** $0.02 \cdot 10^{-6}$

> A *negative* number written in scientific notation is in the form of $a \cdot 10^n$, where n is an integer, and a is a decimal number so that $-10 < a \le -1$. For example, $-7.909 \cdot 10^{10}$ becomes $-79\ 090\ 000\ 000$.

8. Fill in the table.

Scientific Notation	Decimal notation	Scientific Notation	Decimal notation
$-4 \cdot 10^5$		$-7 \cdot 10^{-3}$	
	$-59\ 000\ 000$	$-2.81 \cdot 10^{-7}$	
$-1.506 \cdot 10^6$			-0.0000098
	$-1\ 008\ 200\ 000$		-0.0000000503

9. In 2022, the public debt of the United States was around 26.82 trillion U.S. dollars. India's public debt in 2022 was about 1 261 000 000 000 USD. Write these numbers in scientific notation, as negative numbers. Which country has more debt?

10. Write the numbers in order from smallest to greatest: $4 \cdot 10^{-4}$ $-4 \cdot 10^4$ $-4 \cdot 10^{-4}$ 0.004 $4 \cdot 10^4$

Significant Digits

Significant digits (or figures) of a number are those digits whose value contributes to the precision of the number. They're mainly used in the context of measurements or approximate quantities. (Keep in mind that measurements, by their nature, are never totally precise.)

Example 1. All the individual digits of 12.593 m tell us something about its precision: it is precise to the thousandth of a metre. It has five significant digits.

Example 2. In the measurement 2000 kg, we cannot be sure if the number was originally measured as 1928.34 kg and rounded to 2000 kg, or measured as 2400 kg and rounded to 2000 kg. So in 2000 kg, only the "2" is a significant digit that tells us something about its precision.

Example 3. If someone measured a distance and simply stated it was 410 metres, you won't know whether they measured it to the nearest metre, to the nearest tenth of a metre, to the nearest centimetre, or how. But if they said it was 410.0 metres, then you *would* know they measured it to the nearest tenth of a metre. The number 410 has *two* significant digits, whereas 410.0 has *four*. The latter is much more precise.

Rules for significant digits

All non-zero digits are always significant. With zeros, the situation is more complex. Here are the rules:

1. All non-zero digits are significant: 38.2 has three significant digits.

2. Zeros between other significant digits are also significant: 50 039 has five significant digits.

3. Non-decimal zeros at the end of a number are not significant: 6400 has two significant digits.

4. Decimal zeros in front of the number are not significant: 0.0038 has two significant digits.

5. Decimal zeros at the end of a number *are* significant: 0.00380 has three significant digits.

6. In scientific notation, *all* digits in the decimal part of the number are significant: $5.40 \cdot 10^6$ has three significant digits.

Example 4. To round 4506.90 to three significant digits means using the 4, 5, and one more digit (counting from the left). It means we round to the nearest ten: $4506.90 \approx 4510$.

1. Which measurement is more accurate or precise, 0.506 g or 0.5060 g?

2. How many significant digits do these numbers or quantities have?

a. 24.5	**b.** 20.5	**c.** 24.50	**d.** 0.5
e. 15 000 lb	**f.** 15 001 lb	**g.** 0.078 g	**h.** 0.0780 g
i. $5.0029 \cdot 10^7$	**j.** $3 \cdot 10^7$	**k.** $7.00 \cdot 10^9$	**l.** $5.080 \cdot 10^{-4}$

3. Round each number to the amount of significant figures given in brackets.

a. 2.459 (2)	**b.** 1038.6 (3)	**c.** 29 486 (3)
d. 1736.29 (4)	**e.** 0.302894 (2)	**f.** 1.03287739 (5)
g. 493.2 (1)	**h.** 4 392 342 (3)	**i.** 656 794 000 (4)

In a **calculation involving multiplication and/or division**, the amount of significant digits in the answer should equal the amount of significant digits in the number that is the ***least* precise** (has the least amount of significant digits).

Example 5. How should you round the answer to 24 000 kg ÷ 1.35?

Since 24 000 has two significant digits and 1.35 has three, the answer should be given with two:

$$24\ 000\ \text{kg} \div 1.35 = 17\ 777.\overline{7} \approx \underline{18\ 000\ \text{kg}}$$

However, 24 000 kg *could* be accurate to three significant digits (such as if it was given as $2.40 \cdot 10^4$). Then the answer should be given with three significant digits:

$$24\ 000\ \text{kg} \div 1.35 = 17\ 777.\overline{7} \approx \underline{17\ 800\ \text{kg}}$$

You can use a calculator for all the problems in this lesson.

4. Calculate with a calculator. Round your answer to the correct amount of significant digits.

a. 7.8 m · 2.4 m	**b.** 0.9 m · 7.81 m
c. 630 cm · 45 cm	**d.** 2.4 kg · 51 000
e. $(4.50 \cdot 10^5$ dollars$) \div 130$	**f.** \$5.38 · 45 · 20
g. 3700 km ÷ 4.2 hr	**h.** 49 L ÷ 782 km

5. Calculate an estimation for the total cost of providing 4000 people with a meal costing \$4.59, every day for a year. Assume that 4000 is accurate to two significant digits.

6. The Williams family runs three 0.95 kW (kilowatt) air conditioners for 8.0 hours a day for 30 days. If energy costs 14¢ per kilowatt-hour, calculate the cost of running those AC units for that time. (Treat 30 as accurate to two significant digits.)

   Hint: To get the cost, multiply the price of electricity by the amount of kilowatt-hours. And to get the kilowatt-hours, multiply 0.95kW by the number of hours they run the AC unit in a month.

7. The two sides of a rectangular play area are measured to be 24.5 m and 13.8 m.

   a. Calculate its area and give the result to a reasonable accuracy.

   b. Let's say the dimensions of the play area were measured more accurately to be 24.56 m and 13.89 m. Calculate the area and give the result to a reasonable accuracy.

Measurement conversions involve multiplying or dividing the quantity by a conversion factor. The same rule applies: round your answer to the same amount of significant digits as the number with the least amount of significant digits. In conversion calculations, this is usually the measurement itself, since typically the conversion factors are quite exact and have a lot of significant digits.

8. Convert.

a. 56 in _____ cm

b. 240 cm _____ in

c. 46 m _____ ft

d. 350 gal _____ L

e. 375 L _____ gal

f. 125 ft _____ m

1 inch = 2.54 cm

1 ft = 0.3048 m

1 gal = 3.785 L

In a calculation involving **addition and/or subtraction**, we don't use significant digits when rounding the answer. Instead, we look at the places and round the answer to the **same accuracy (same number of decimal places)** as the number that is the least accurate (with the fewest decimal places). The answer cannot be any more precise than the least precise measurement that went into it.

Example 6. In calculating 9.12 m + 7 m + 4.5 m, the quantity 7 m has zero decimal places and is the least accurate, so the final answer should be rounded to the nearest meter.

9. Calculate, and give your answer to a reasonable accuracy.

a. 5.6 kg + 2.04 kg − 0.078 kg ≈ _____ kg

b. 7.6 m + 0.752 m + 2.09 m ≈ _____ m

c. 14 lb + 7.8 lb + 55 lb ≈ _____ lb

d. 506 mi + 78 mi + 5.9 mi ≈ _____ mi

Example 7. In reality, how long is a distance of 5.7 cm on a map with a scale of 1:400 000?

Since 1 cm corresponds to 400 000 cm, then 5.7 cm corresponds to 5.7 · 400 000 cm = 2 280 000 cm. Converting this into kilometres we get 22.8 km.

The scale ratio 400 000 is accurate; it is not 399 999 nor 401 000, but exactly 400 000. So it has six significant digits. The distance 5.7 has two, so the final result needs to have two, and be given as 23 km.

10. Calculate the following distances in reality.

   a. 6.2 cm on a map with a scale of 1:50 000

   b. 12.5 cm on a map with a scale of 1:200 000

11. The distance from Mary's home to school is 7.6 cm on a map with a scale of 1:10 000. How long is this distance in reality, in metres?

12. A field measures 5.0 cm by 3.5 cm on a map with a scale of 1:8000. Calculate its area in reality.

Using Scientific Notation in Calculations, Part 1

Example 1. How many times bigger is one number than another?

You can easily tell that $60 is three times as much as $20. But what about $500 000 and $150 000 000? Scientific notation makes these types of comparisons very straightforward.

First we write the numbers in scientific notation: $\$500\,000 = 5 \cdot 10^5$ and $\$150\,000\,000 = 1.5 \cdot 10^8$. Next we

divide them, using the quotient rule for exponents: $\dfrac{1.5 \cdot 10^8}{5 \cdot 10^5} = \dfrac{1.5}{5} \cdot \dfrac{10^8}{10^5} = 0.3 \cdot 10^3 = 0.3 \cdot 1000 = 300$.

So, the larger number is 300 times the other. No calculator needed, and in fact, if the exponents had been larger, a regular calculator would not handle the numbers in decimal notation.

Don't confuse the above with simple comparisons where we determine which number is greater, such as $32\,000 < 6 \cdot 10^4$. The above is a *multiplicative* comparison: how many *times* bigger is one number than another?

Do not use a calculator in the problems on this page.

1. The mass of the sun is about $2 \cdot 10^{30}$ kg. The mass of the Earth is about $6 \cdot 10^{24}$ kg. About how many times more massive is the sun than the earth?

2. **a.** How many times bigger is $6 \cdot 10^{-20}$ than $3 \cdot 10^{-30}$?

   b. How many times bigger is $2 \cdot 10^4$ than $8 \cdot 10^{-4}$?

3. The speed of light is approximately $3 \cdot 10^5$ km/s. The distance from earth to sun is approximately 150 million kilometres.

   a. Write the distance in scientific notation.

   b. Now use the two numbers that are in scientific notation, and calculate how long it takes for sunlight to travel from the sun to the earth.

   Give thought to *which* unit of time you will use for the answer; in other words, which unit makes most sense considering the context.

Example 2. When Sheila calculated the value of 70^{23} with a calculator, she got this:

> 70 ^ 23 =
>
> **2.73687473400809l6343e+42**

The number is given in scientific notation, signifying $2.7368747340080916343 \cdot 10^{42}$. (In reality, the decimal digits would continue for longer; we only see the digits that fit on the calculator screen.)

The letter "e" refers to "exponent"; however this is not referring to the decimal number being raised to the power of 42, but the <u>number 10</u> being raised to that power.

You may use a calculator for all the rest of the problems in this lesson.

4. A student multiplied two large numbers with a calculator and got this: 1.5E26

   a. What does the answer mean?

   b. What two numbers could she have multiplied?

5. In scientific notation, we use negative exponents for numbers with very small absolute value. Investigate how different calculators show this. *Hint*: divide a small number by a very large number.

6. The speed of light is 299 792 458 m/s. Calculate the distance light travels in a year. This distance is called a *light year*. Give your answer in kilometres, in scientific notation, and with four significant digits.

7. A scientific paper from 2016 estimates that an average 70-kg man has about $3.8 \cdot 10^{13}$ bacteria in his body (most are gut bacteria), and that those bacteria have a mass of about 0.2 kg. What is the average mass of one bacterium in this scenario? (Round your answer considering the significant digits.)

8. A golden eagle can dive at a speed of $2.10 \cdot 10^7$ cm per hour. A garden snail is 4600 times slower than the eagle! Find the speed of the garden snail and give it in a reasonable unit, and considering significant digits.

Example 3. The mass of one gold atom is about $3.2696 \cdot 10^{-22}$ grams. How many gold atoms are there in one troy ounce of gold? (1 troy ounce = 31.10348 g)

This is a division problem. We divide 31.10348 grams by the mass of one gold atom: $\dfrac{31.10348 \text{ g}}{3.2696 \cdot 10^{-22} \text{ g}}$.

First off, note that the units "g" cancel out, which is what we would expect, since we expect to get a number without any units (a quantity or "how many").

We will write this quotient in two parts, as $\dfrac{31.10348}{3.2696} \cdot \dfrac{1}{10^{-22}}$, and then work with the two parts separately.

From the calculator, $\dfrac{31.10348}{3.2696} \approx 9.5129$ (five significant digits). The other part, $\dfrac{1}{10^{-22}}$, equals 10^{22}.

The end result is that you need about $9.5129 \cdot 10^{22}$ gold atoms to make one troy ounce of gold.

9. The mass of one gold atom is about $3.2696 \cdot 10^{-22}$ grams.

   a. What is the approximate mass of a trillion gold atoms?

   b. Use the table on the right and give this mass using an appropriate prefix with the unit "gram". For example, the mass of $5 \cdot 10^{-7}$ grams could be given as 0.5 micrograms or as 500 nanograms.

Prefix	Meaning
milli	10^{-3}
micro	10^{-6}
nano	10^{-9}
pico	10^{-12}
femto	10^{-15}
atto	10^{-18}

10. Recall that the nucleus of an atom consists of protons and neutrons, and electrons are very small particles that whiz around the nucleus.

    We commonly see images like this, where it looks like the nucleus is maybe about 1/4 of the diameter of the entire atom. But what is the truth of the matter?

    Let's look at silicon, for example. The radius of a silicon atom is about 110 picometres. The radius of the *nucleus* of a silicon atom is about 3.6 femtometres.

    In the case of silicon, about how many times bigger is the diameter of the entire atom than the diameter of the nucleus?

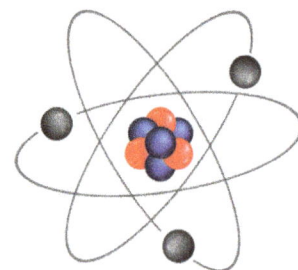

Using Scientific Notation in Calculations, Part 2

1. Some students get confused with the rules of exponents when *adding* numbers in scientific notation. Compare the problems carefully, and solve. Give your answers in decimal notation. Do not use a calculator.

   a. $(2 \cdot 10^6) \cdot (3 \cdot 10^4)$

   b. $2 \cdot 10^6 + 3 \cdot 10^4$

   c. $8 \cdot 10^3 + 7 \cdot 10^5$

   d. $(8 \cdot 10^3) \cdot (7 \cdot 10^5)$

   > It is easy to add or subtract numbers in scientific notation IF they have the same power of ten: you can simply add or subtract their decimal parts.
   >
   > **Example 1.** Add $2.81 \cdot 10^{13} + 5.2 \cdot 10^{12}$.
   >
   > We will write $2.81 \cdot 10^{13}$ with 10^{12} instead of 10^{13}: $2.81 \cdot 10^{13} = 28.1 \cdot 10^{12}$. Now, the problem becomes $28.1 \cdot 10^{12} + 5.2 \cdot 10^{12}$. We can simply add $28.1 + 5.2 = 33.3$. The final sum is $33.3 \cdot 10^{12}$.
   >
   > Another possibility is to simply use decimal notation to add the numbers, like you probably did in question #1. This works if the absolute values of the exponents are not very large.

2. Solve. Give your answer in scientific notation.

a. $4.8 \cdot 10^8 + 5 \cdot 10^7$	b. $9.3 \cdot 10^6 + 8 \cdot 10^7$	c. $5 \cdot 10^7 - 7 \cdot 10^5$	d. $8.4 \cdot 10^9 - 4.7 \cdot 10^8$

3. Jeremy and Mia were working on the problem $5 \cdot 10^{-3} + 2 \cdot 10^{-4}$. One of them got the answer $7 \cdot 10^{-7}$ and the other got $5.2 \cdot 10^{-4}$. Is either answer correct? If not, find the correct answer (without a calculator).

4. Solve.

a. $8 \cdot 10^{-2} + 6 \cdot 10^{-3}$	b. $3 \cdot 10^{-6} + 5 \cdot 10^{-5}$
c. $2 \cdot 10^{-4} - 7 \cdot 10^{-6}$	d. $5.4 \cdot 10^{-3} - 7 \cdot 10^{-4}$

5. Compare the volumes of different planets, the moon, and the sun that are given in the table, in cubic kilometres.

Celestial Body	Volume
Earth	$1.0832 \cdot 10^{12}$ km³
Moon	$2.1968 \cdot 10^{10}$ km³
Mars	$1.6318 \cdot 10^{11}$ km³
Jupiter	$1.4313 \cdot 10^{15}$ km³
Sun	$1.4093 \cdot 10^{18}$ km³

   a. How many Jupiters would "fit" in the sun?

   b. How much bigger is the volume of the earth than Mars?

   c. What is the combined volume of the earth and the moon?

6. Count how many breaths you take in a minute (at rest), and from that, estimate how many breaths in total you would take in a 70-year lifespan. Then choose the closest estimate from the options below.

   The number of breaths a person takes in a lifetime is about: **a.** $5 \cdot 10^6$ **b.** $5 \cdot 10^8$ **c.** $5 \cdot 10^{10}$ **d.** $5 \cdot 10^{12}$

7. A 70-kg male body contains approximately $7 \cdot 10^{27}$ atoms. Of these, approximately $4.22 \cdot 10^{27}$ are hydrogen atoms, $1.61 \cdot 10^{27}$ are oxygen atoms, $8.03 \cdot 10^{26}$ are carbon atoms, and $3.9 \cdot 10^{25}$ are nitrogen atoms.

   a. About what percentage of the atoms in a human body are hydrogen atoms?

   b. About how many more times oxygen atoms does the human body have than nitrogen atoms?

   c. About how many more oxygen atoms does the human body have than carbon atoms?

8. It is estimated that there are about 10^{15} ants on this planet, and that the average mass of each ant is about 1 mg.

   a. Find the total mass of the ants living on this planet. Give your answer in a sensible unit.

   b. Now find the total mass of humans living in Asia, in kilograms. Use 60 kg for the average weight of humans in Asia, and 4800 million for the population of Asia (or check the current population at https://www.worldometers.info/world-population/asia-population/).

   c. Which have a larger mass, all the ants on the planet, or the people living in Asia?

9. The water volume in Lake Victoria is approximately 2750 km^3.

   a. Convert this to cubic metres and write the resulting number in scientific notation.
   Hint: the unit "km^3" means "1000 metres, cubed", or $(1000 \text{ m})^3$.

   b. Now calculate how many bucketfuls of water there are in Lake Victoria. Use 20 litres for the volume of one bucket (consider it accurate to two significant digits). (1 m^3 = 1000 litres.)

Chapter 1 Review

1. Find the value of the expressions.

a. $(-2)^4 =$	**b.** $-2^4 =$	**c.** $8^{-2} =$	**d.** $5^2 \cdot 5^8 \cdot 5^{-7} =$
e. $11 \cdot 10^{-2} =$	**f.** $10^3 + 10^4 =$	**g.** $\left(\dfrac{2}{-3}\right)^3 =$	**h.** $\dfrac{12^7}{12^5} =$

2. Write an equivalent expression using the exponent laws, without negative exponents.

a. $(a^{-1})^4 =$	**b.** $(2x)^3 =$	**c.** $(5x)^{-2}$	**d.** $-2s^5 t^7 t^3 \cdot 4s^8 =$
e. $\dfrac{9a^7}{30a^5} =$	**f.** $\dfrac{x^3}{x^{-2}} =$	**g.** $\left(\dfrac{3x}{-4}\right)^3 =$	**h.** $\left(\dfrac{2a^2}{b}\right)^5 =$

3. Find the value of each unknown.

a. $8^x 8^5 = 8^{24}$	**b.** $(7^8)^{-3} = \dfrac{1}{7^y}$	**c.** $\left(\dfrac{2}{3}\right)^x = \dfrac{16}{81}$	**d.** $\dfrac{(-3)^x}{(-3)^5} = -27$	**e.** $(3^z)^2 = 9^4$

4. Find the true statements.

a. $(a \cdot b)^2 = a^2 \cdot b^2$	**b.** $a^m \cdot a^n = a^{m+n}$	**c.** $(a \cdot b)^{-3} = \dfrac{1}{ab^3}$	**d.** $(a+b)^2 = a^2 + b^2$

5. Natalie wrote the prime factorization of 21 600. Find the error.

$$21\ 600$$
$$= 100 \cdot 216$$
$$= 10^2 \cdot 6^3$$
$$= (2 \cdot 5)^2 \cdot (2 \cdot 3)^3$$
$$= 2^2 \cdot 5^2 \cdot 2^2 \cdot 3^3$$
$$= 2^4 \cdot 3^3 \cdot 5^2$$

6. There is one pretty simple number x for which the expression $(x + 1)^3$ is equal to $x^3 + 1$. Find that number.

7. Calculate. Give each answer to a reasonable accuracy.

a. 164.3 km $\div 2.5$ hr	**b.** 3.6 m $+ 89$ m $+ 0.3$ m	**c.** $5.210 \cdot 10^9$ dollars $\div 365$

8. Calculate. Give each answer to a reasonable accuracy.

   a. Sydney jogs through a 3.5-km jogging track twice every week.
      What is the total distance she jogs in a year?

   b. Your car's odometer shows you travelled 998.4 km since you last filled your fuel tank. To fill
      your tank this time, it took 81.9 litres of gas. What is your fuel consumption, in litres per 100 km?
      Hint: divide the litres by (kilometres/100).

   c. What is the area of a rectangle with 2.3 m and 11.9 m sides?

9. Complete the chart by rewriting each distance in scientific notation.

Planet	Average distance from sun (km)	In scientific notation (km)
Mercury	58 000 000	
Jupiter	778 570 000	
Neptune	4 495 000 000	

10. Radioactive elements undergo radioactive decay, where they lose energy by
    radiation. Half-life is the time required for exactly half of these entities to
    decay, on average.

    The half-life of thorium-217 is 240 microseconds, and the half-life of uranium-216
    is 4.3 milliseconds.

Prefix	Meaning
milli	10^{-3}
micro	10^{-6}
nano	10^{-9}

    a. Write these two amounts in scientific notation, in *seconds*.

    b. How much longer is the half-life of uranium-216 than the half-life of
       thorium-217? Give your answer in milliseconds.

11. The average speed of a garden snail is about $1.3 \cdot 10^{-2}$ m/s. The average speed of a cheetah is about
    27.7 m/s. About how many times faster is the cheetah than the garden snail?

48

Chapter 2: Geometry
Introduction

The second chapter of Math Mammoth Grade 8 covers geometric transformations, angle relationships, and the volume of prisms, cylinders, pyramids, cones, and spheres.

The chapter starts out with the basics of congruent transformations: translations, reflections, rotations. Students use transparent paper to perform several of these transformations hands-on, so as to gain an understanding of the attributes that are preserved in these transformations.

Next we practise these same transformations in the coordinate grid. Students learn how the coordinates of the points change when a figure is translated or reflected in the *x* or *y*-axis. They also explore rotating figures in the coordinate grid; here we limit the rotations to 90°, 180°, or 270° degrees.

Then it is time to study sequences of transformations, which enable us to describe more complex transformations. The key idea here is to understand that a two-dimensional figure is congruent to another if the second can be obtained from the first by a sequence of transformations.

All of this work has related to congruent transformations, which means the size of the figure has not changed. Now we turn our attention to dilations. In a dilation, the figure is transformed so that its size changes but its shape does not. Such figures are called similar figures. Yet another term describing the same process is scaling a figure.

Next, we study angle relationships. The first lesson in this section reviews certain angle relationships from 7th grade (complementary, supplementary, and vertical angles). Then students learn about angles formed when a transversal crosses two parallel lines: corresponding angles, alternate interior angles, and alternate exterior angles. They also investigate angle relationships related to triangles and learn how these relationships allow us to deduce angle measurements of other angles.

In all of this work, students are guided to reason using mathematical facts they have learned, and to justify their reasoning, thus becoming familiar with the process of mathematical proof.

The last major topic of the chapter is volume of various three-dimensional figures. Students solve a variety of real-world and mathematical problems involving multiple three-dimensional shapes.

Pacing Suggestion for Chapter 2

This table does not include the chapter test as it is found in a different book (or file).
Please add one day to the pacing if you use the test.

The Lessons in Chapter 2	page	span	suggested pacing	your pacing
Geometric Transformations and Congruence, Part 1	51	*4 pages*	1 day	
Geometric Transformations and Congruence, Part 2	55	*3 pages*	1 day	
Translations in the Coordinate Grid	58	*3 pages*	1 day	
Reflections in the Coordinate Grid	61	*3 pages*	1 day	
Translations and Reflections	64	*3 pages*	1 day	
Rotations in the Coordinate Grid	67	*4 pages*	1 day	
Sequences of Transformations	71	*3 pages*	1 day	
Sequences of Transformations, Part 2	74	*2 pages*	1 day	
Dilations	76	*3 pages*	1 day	
Dilations in the Coordinate Grid	79	*3 pages*	1 day	

Helpful Resources on the Internet

We have compiled a list of Internet resources that match the topics in this chapter, including pages that offer:

- **online practice** for concepts;
- online **games**, or occasionally, printable games;
- **animations** and interactive **illustrations** of math concepts;
- **articles** that teach a math concept.

We heartily recommend you take a look! Many of our customers love using these resources to supplement the bookwork. You can use these resources as you see fit for extra practice, to illustrate a concept better and even just for some fun. Enjoy!

https://l.mathmammoth.com/gr8ch2

Scan me

Geometric Transformations and Congruence, Part 1

Two figures are congruent when they are, you might say, identical in the sense that they have the same shape and size (but may be of different colour). We can define congruency as follows:

> Two figures are **congruent** if they perfectly match, when one is placed on top of the other.

The figures don't have to be in the same position or orientation. For example, these two figures are congruent — if you rotate and move figure A, you can place it exactly on top of figure B.

FIGURE A

FIGURE B

We will now study three geometric **transformations**, or basic ways to move a point, or by extension, a figure, since a figure can be considered to consist of many points.

1. A **translation** of a figure means sliding or moving it a certain distance in a certain direction, without turning or rotating it. The arrows show how three individual points of the figure were moved.

   We say the translation maps point A onto point A' (read "A prime"), point B onto point B', and point C onto point C'.

   We also say that point A' is the image of point A under the translation.

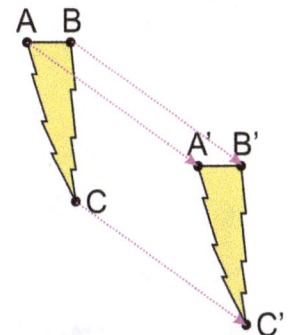

2. A **rotation** means turning a figure around a certain point.

   Here, the lightning figure is rotated around point P. Each point of the figure moves in a **circular arc around point P**.

   A rotation is measured in degrees, just like angles are. In this example, the lightning figure was rotated 67 degrees clockwise around point P.

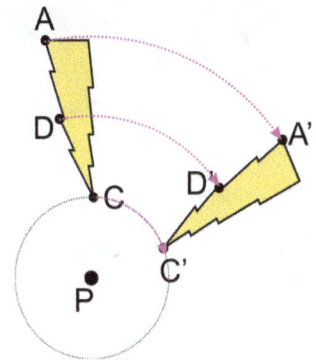

3. A **reflection** across a line means mirroring the figure in that line. You could also say the figure was "flipped".

   In a reflection, the distance from each point to the reflection line and the distance of its image to the line are equal (measured along a line segment that is perpendicular to the line).

   For example, the distance from point C to the line equals the distance from point C' to the line.

   A reflected figure is congruent to the original.

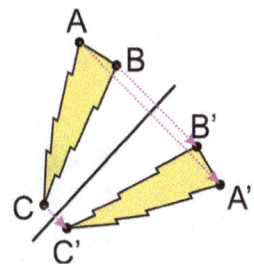

1. Name the transformation that was used to transform the figure on the left to the figure on the right.

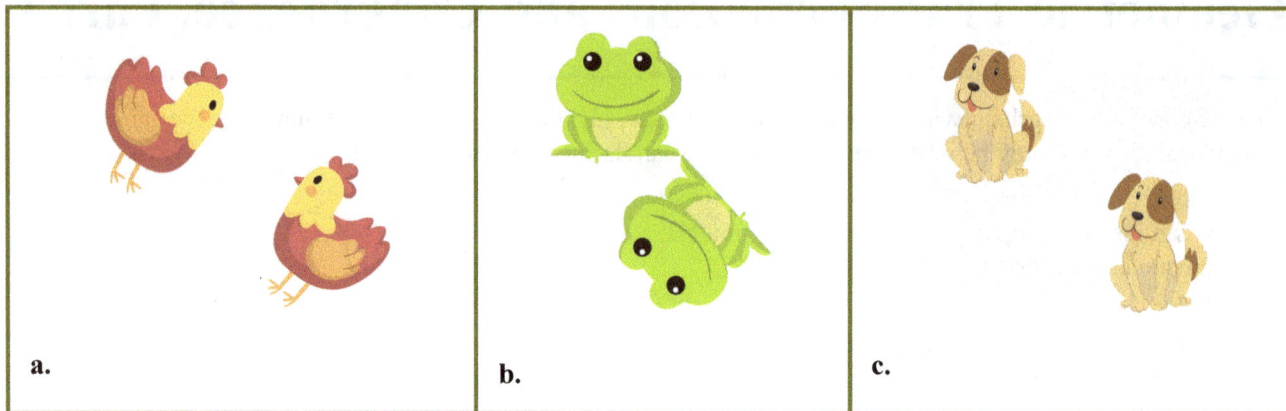

a.

b.

c.

In continuation, we will explore geometric transformations and how they relate to congruence with the help of tracing paper (patty paper) or a transparency.

2. Use tracing paper to determine whether the two figures are congruent. You may move, turn, and/or flip the tracing paper. First, copy the outline of **one** figure to the tracing paper.
   (Note: when checking for congruency, we ignore the colours.)

a.

b.

c.

3. The image below shows how point A was mapped to point A' in a rotation. We will now do the same rotation to points B and C using tracing paper. This is how:

i. Put a thumbtack or a pin through the tracing paper at P so that you can turn the paper around P.
ii. Copy points A, B, and C to the paper.
iii. Then rotate the paper around point P so that **point A is mapped to point A'**.
iv. Now, draw the points B' and C'. You can use a pin to mark where these points are (through the tracing paper). Drawing the points with a pencil on the tracing paper may also make a faint mark in the underlying paper. Then remove the tracing paper and draw the points.

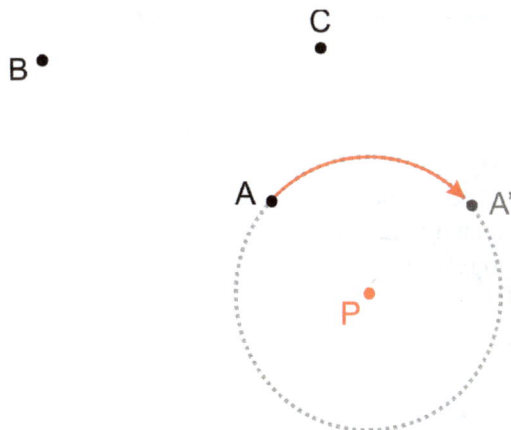

a. Connect A, B, and C with line segments, and also A', B', and C', so that you get two triangles.

b. Measure the side lengths of both triangles. What do you notice?

c. Measure the angles BAC and B'A'C' and also the angles ACB and A'C'B'. What do you notice?

4. Point X' is the image of point X under a translation along the dashed arrow.

a. Sketch the image of point Y in the same translation. Mark it as point Y'.

You may optionally do this translation with tracing paper. However, it is difficult to do this accurately.

b. What can we know about the length of the segment $\overline{X'Y'}$? Choose one answer:

(i) $\overline{XY}$ and $\overline{X'Y'}$ are congruent (have the same length).

(ii) $\overline{XY}$ and $\overline{X'Y}$ are not congruent.

(iii) We cannot know for sure whether $\overline{XY}$ and $\overline{X'Y}$ are congruent or not.

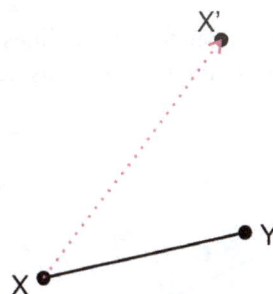

How to reflect a point across a line using tracing paper or a transparency	
Step 1. Align the paper so that one of its edges is along the reflection line m.	**Step 2.** Flip the paper. You can use a pin to mark the image of the point in question.

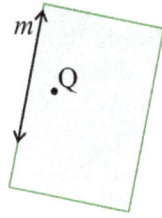

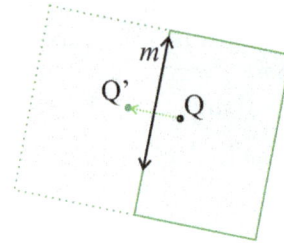

5. **a.** Cut out a piece of transparent paper that fits inside the light-coloured rectangle in the image on the right (approximately 3.2 cm by 4.8 cm). Use tracing paper to reflect the points Q, R, and S across line n. Label the reflected points as Q', R', and S'.

   b. Connect the points Q and R, R and S, Q' and R', and R' and S' with line segments.

   c. Measure the length of the line segments $\overline{QR}$ and $\overline{Q'R'}$, and also $\overline{RS}$ and $\overline{R'S'}$. What do you notice?

   d. Measure also the angles ∠QRS and ∠Q'R'S'. What do you notice?

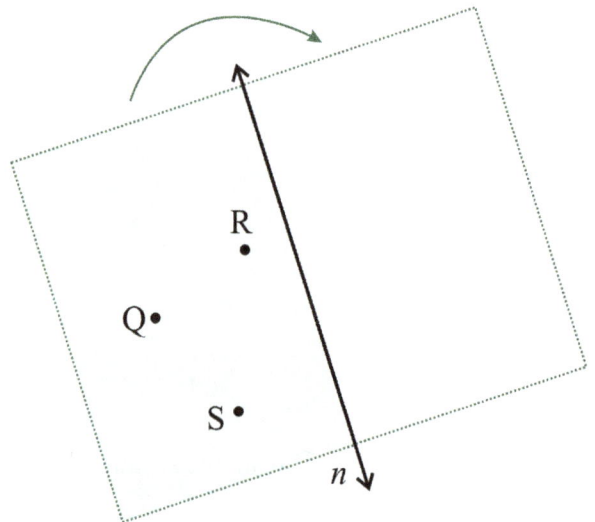

6. Predict what will happen to parallel lines under translation, rotation, and reflection. You may want to use tracing paper (as needed) to confirm your prediction.

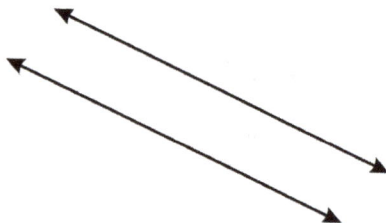

Geometric Transformations and Congruence, Part 2

Rigid transformations

You observed during the previous lesson that translations, reflections, and rotations **preserve lengths, angles, and parallel lines.**

That is why we also call them **rigid transformations** (or isometries): they treat figures in a "rigid" manner, without distorting them. However, they do not preserve the position of the transformed figures.

The image of a figure under these transformations is **congruent** to the original (has the same size and shape.)

A note on notation / symbols. The symbol ∠ signifies an angle and △ signifies a triangle.
So, ∠DEF means angle DEF and △ABC means triangle ABC.

1. Quadrilateral DEFG is reflected across line *t*.
   Certain side lengths are marked in the figure.

   Properly label the vertices of the reflected figure,
   and find its perimeter.

   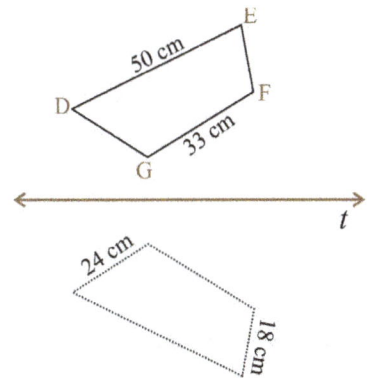

2. A pentagon is rotated around one of its vertices, then reflected.
   Is the resulting pentagon congruent to the original?

   How can we know that?

   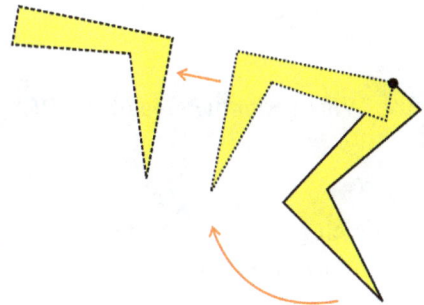

3. Trapezoid DEFG undergoes a reflection, and then a rotation around point P.

   a. Properly label the vertices of the reflected
   figure (with D', E', F', and G'), and of the
   rotated figure (with D", E", F", and G").

   b. Which of the attributes of the trapezoid
   stay the same? Tick all that apply.

   (i) Perimeter

   (ii) Area

   (iii) Position

   (iv) Angle DEF

   (v) Angle sum

   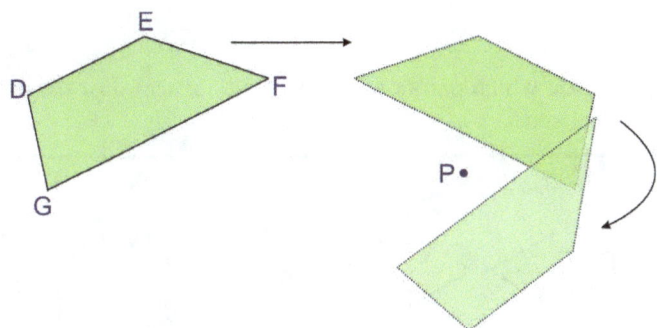

4. Jeannie says that the figure on the right was rotated 180 degrees around point Q to produce the figure on the left.

   Matthew says it is a reflection in the line *l*.

   Who is correct?

   Kim says she can also do it by using translations. Is she right?

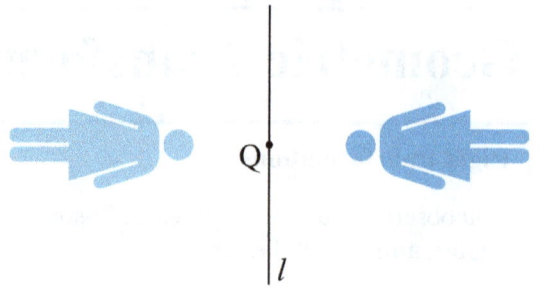

5. Triangle ABC is rotated around point P so that point A maps to point A'.

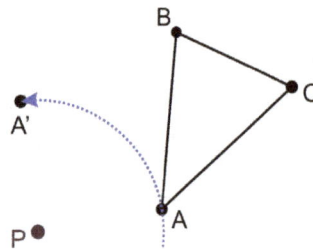

   a. Use transparent paper to draw the image of △ABC under this rotation.

   b. Will the angle measures of △ABC change?

   c. What about the area of △ABC?

6. Which geometric transformation was used to create this image?

7. Decide which pairs of figures are congruent. For the congruent ones, write down the transformation that was used.

   a.

   b.

   c.

   d.

8. Which of the umbrellas, A, B, C, or D is an image of the umbrella on the left under rotation around point P? Use transparent paper to check.

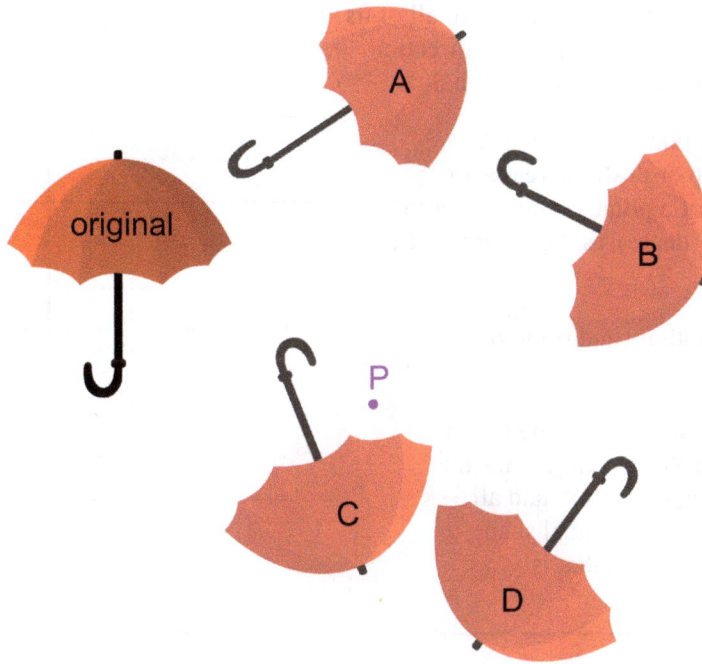

9. Jo thinks that figure B is congruent to figure A since it is a reflection of figure A across the line *S*.

   Is she correct in her thinking?

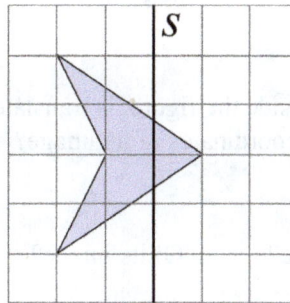

Figure A

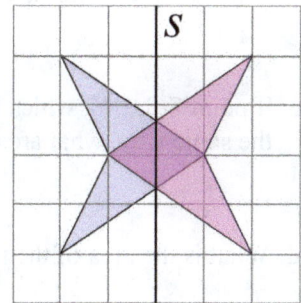

Figures A and B

10. a. Reflect the shape in the dashed line.

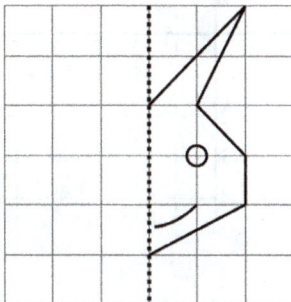

    b. Design your own shape and reflect it.

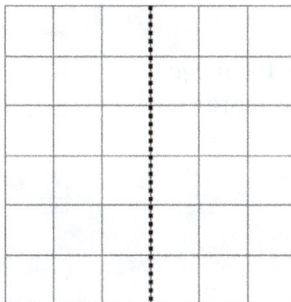

    c. Design your own shape and reflect it.

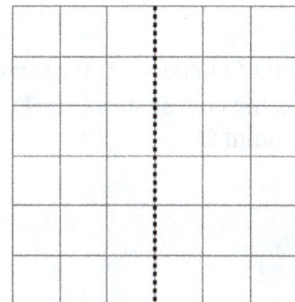

Translations in the Coordinate Grid

When a figure is translated in the coordinate grid, all of its points move the same distance and in the same direction, and the resulting figure is congruent to the original.

Notice how the movements relate to the actual coordinates: a movement up or down increases or decreases the *y*-coordinate of a point. A movement to the right or left increases or decreases the *x*-coordinate of a point.

(What if the movement is both up and right, or both down and left?)

By now we have seen that when translating a shape, we end up with a congruent shape. This means that the side lengths, angles, area, perimeter, and all other geometric properties of the original shape remain the same. The only attribute that is not preserved is location.

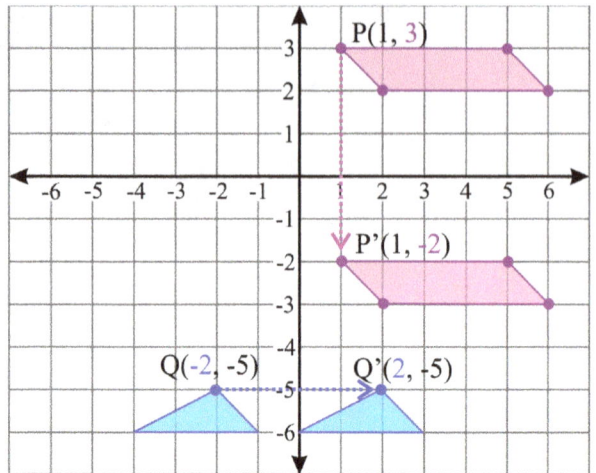

1. **a.** Translate the figure four units up and seven units to the left. Label the image of point A as A'.

   b. What are the coordinates of point A and of point A'?

   c. If point S(3, −4), which is inside the figure, is translated the same way, what are the coordinates of its image?

   d. What is the area of the figure?   Of its image?

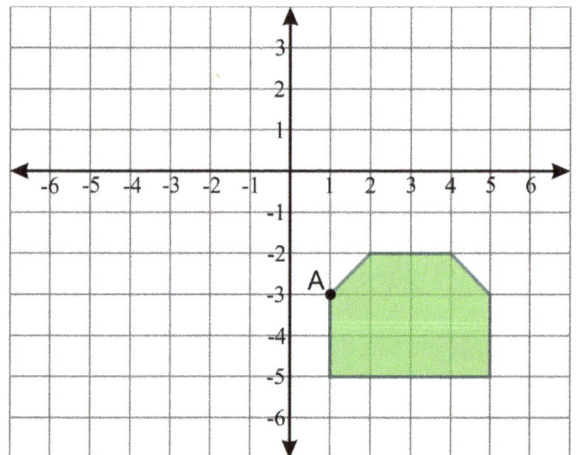

2. **a.** Point P(−4, 3) is translated eight units to the right and six units down. What are the coordinates of its image, P'?

   b. Point Q underwent the same translation. Its *image* Q' has the coordinates (5, −4). What are the coordinates of point Q?

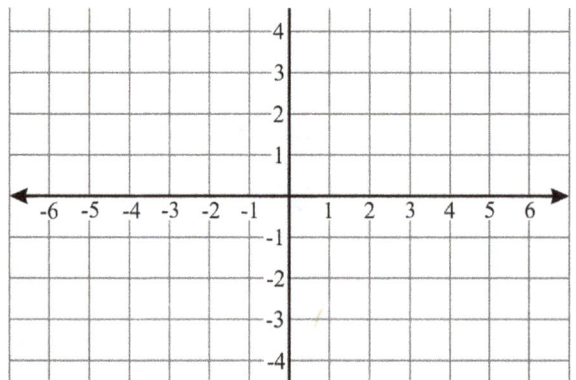

3. Jane claims that △ABC is congruent to △A'B'C' because △ABC could be translated six units to the right and four units up to map onto △A'B'C'.

   Is Jane correct? Explain.

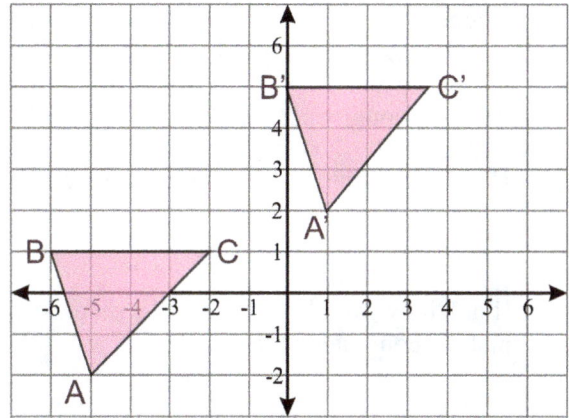

4. You want to prove that figure F' is congruent to figure F by using a translation. How would you do it? Be specific in your explanation, and describe the transformation(s) needed accurately, using the coordinate plane as a reference.

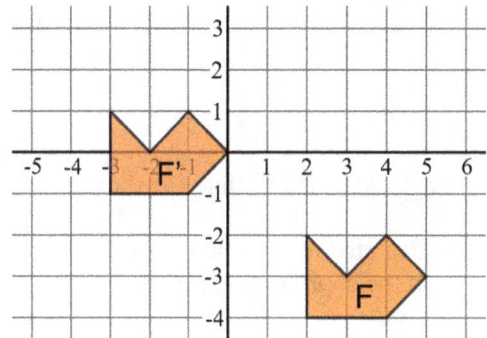

5. The side lengths of a rectangle are 20 and 35 units. It is then translated 50 units down and 15 units to the left.

   a. What is the perimeter of the *translated* rectangle?

   b. How do you know?

6. a. What type of quadrilateral is figure ABCD?

   b. Figure ABCD is translated five units to the left, and two up, to become A'B'C'D'. Name two line segments that are parallel to $\overline{AD}$ in the translated figure.

   c. Angle BAD is 105°. What is the measure of ∠B'A'D'?

   d. Name two features of quadrilateral ABCD that are preserved in the translation, and one that is not.

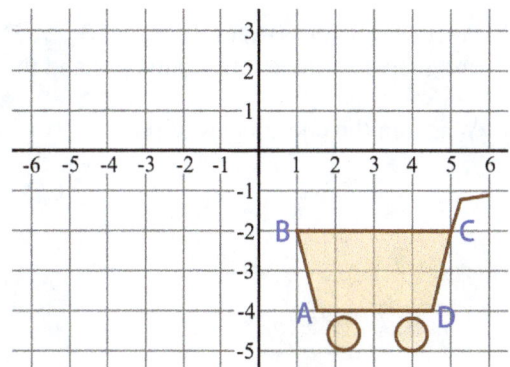

7. The coordinates of triangle 1 are (6, 2), (6, 4), and (3, 5).
   The coordinates of triangle 2 are (−2, 1), (−2, 3), and (−6, 4).

   a. Are they congruent?

   b. If yes, what translation could be used to map one onto
   the other?

   If not, how would you change the second triangle to
   make it congruent to the first?

   c. Calculate the areas of (original) triangles 1 and 2.
   Are they equal?

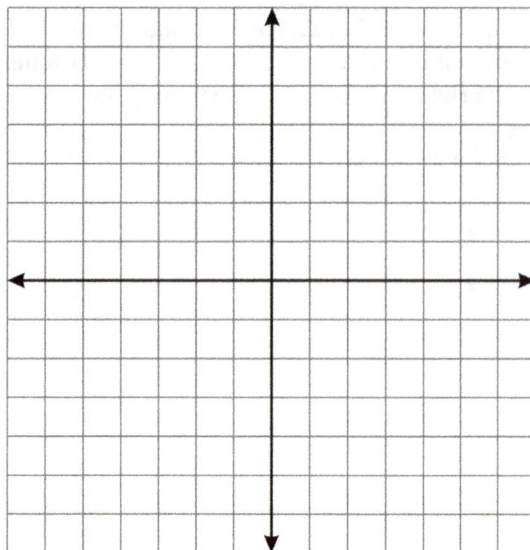

8. Angle ABC is translated as shown.

   a. Describe the translation (in terms of the units
   in the coordinate grid).

   b. Do the lines $\overleftrightarrow{BC}$ and $\overleftrightarrow{B'C'}$ ever intersect?

   c. How do you know?

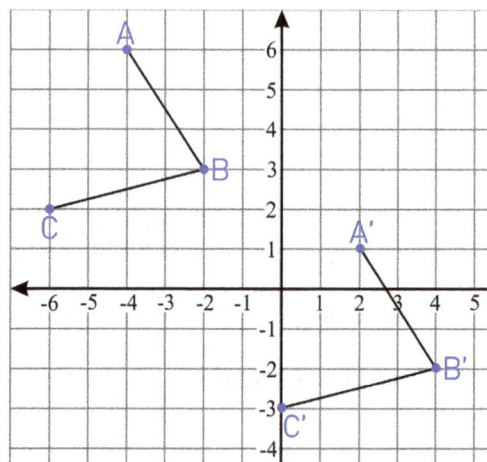

9. A triangle with vertices A(−8, 5), B(−5, 3), and C(−4, 1) is translated six units to the right and two down,
   and then also three units to the left and three down.

   What are the coordinates of point C'' (the image of point C under the double translation)?

Puzzle Corner

Two vertices of triangle ABC are A(1, −6) and B(1, −2),
and its area is 12 square units.
What are the coordinates of point C?

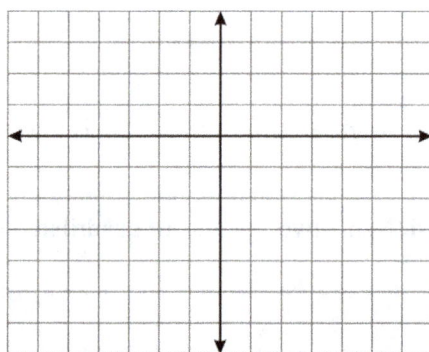

60

Reflections in the Coordinate Grid

To reflect point P across line *l*, draw a line segment from point P that is perpendicular* to line *l*. Continue the line segment. The reflected point P' is at the same distance from line *l* as P, just on the other side.

In other words, in a reflection, each point and its image are at an equal distance from the line of reflection, measured along a line that is perpendicular to the line of reflection.

*Two lines or line segments are **perpendicular** if they meet at a right angle.

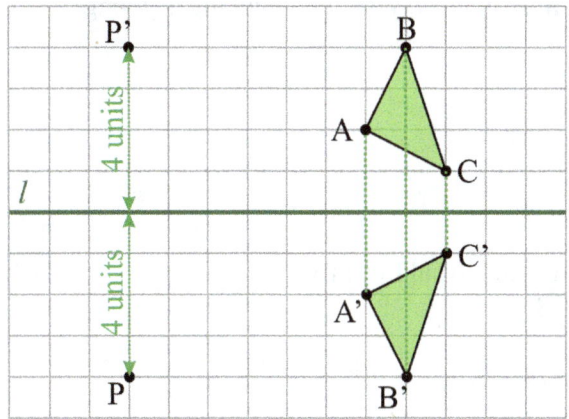

1. **a.** Reflect the points across line *s*.

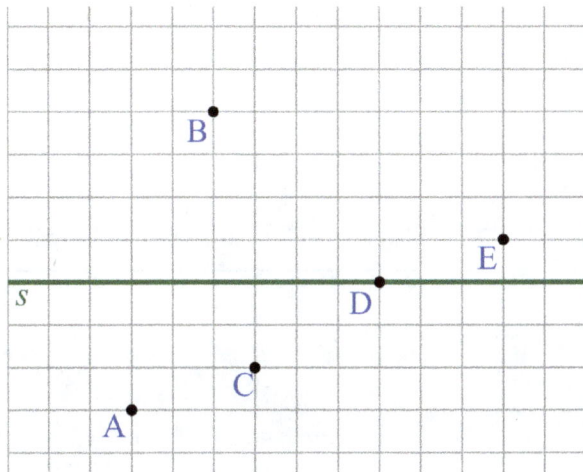

b. Reflect the figures across line *t*.

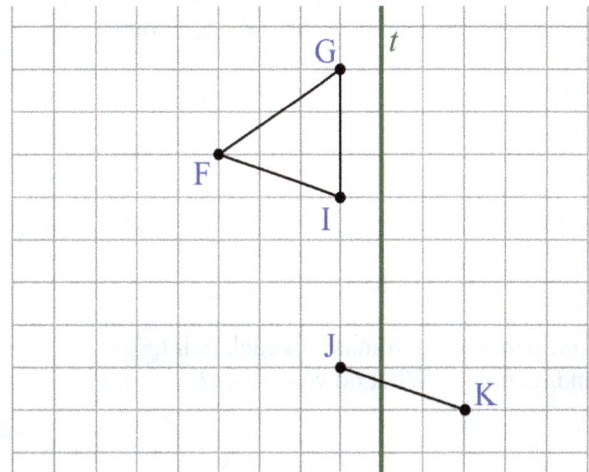

2. **a.** Draw a vertical line that passes through the point (2, 0).

b. Draw the points P(1, 2), R(3, 1), and Q(5, 4).

c. Reflect each point across the line. Label the reflected points as P', R', and Q'.

d. Lastly, connect P, Q, and R to form a triangle, and also P', Q', and R'.

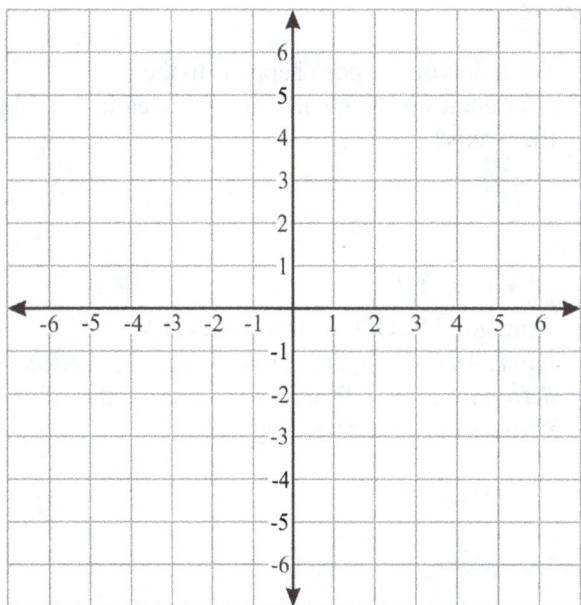

3. James says that figure 2 is congruent to figure 1 because it is a reflection of figure 1 across the horizontal line L.

   a. Explain why James's thinking is wrong.

   b. How would you fix the situation?

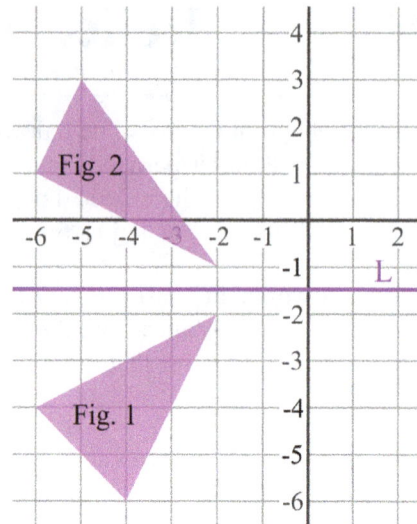

Fig. 2

Fig. 1

L

4. Reflect the points listed below in the *x*-axis. Write down the coordinates of the reflected points:

   H (−2, 3) → H' (_____ , _____)

   I (1, −1) → I' (_____ , _____)

   J (3 , 5) → J' (_____ , _____)

   K (−5 , −4) → K' (_____ , _____)

   Compare the coordinates of each point and its image. What do you notice?

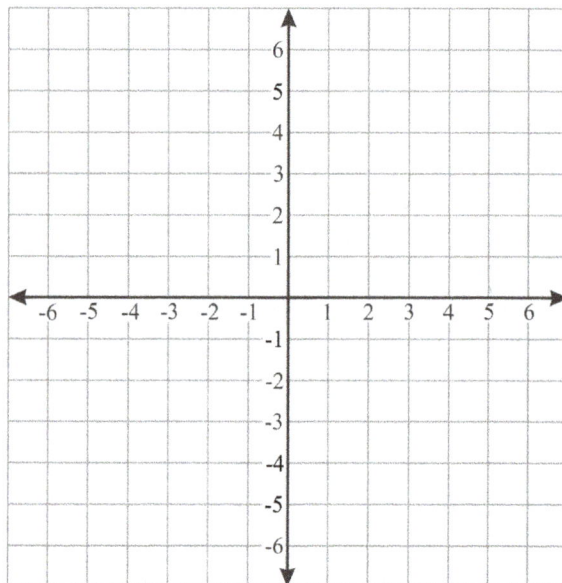

   What do you suppose happens to the coordinates of points that are reflected in the *y*-axis?

5. Pentagon MNOPQ with vertices at M(−3, 1), N(−1, 4), O(3, 4), P(5, 1), and Q(0, −1) is reflected across the *x*-axis. What are the coordinates of the vertices of the reflected figure?

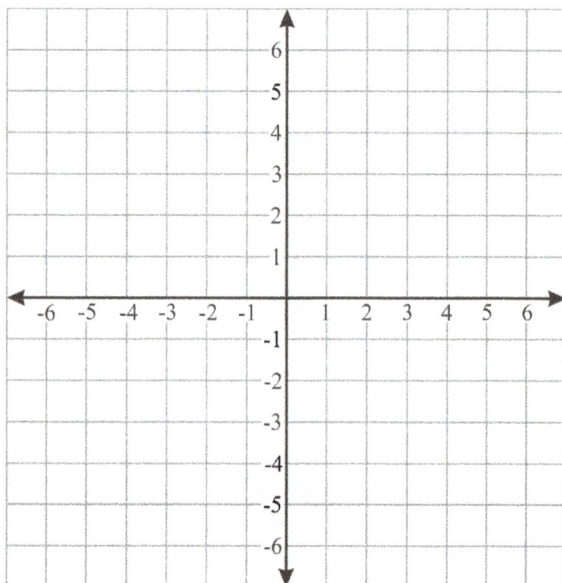

6. Look at the various figures and their reflections in this lesson, and also think back to the earlier lessons on congruence. Which attributes are preserved in a reflection of a figure?

   a. perimeter   **b.** orientation   **c.** location   **d.** area   **e.** measure of angles

7. In each image, a figure has been reflected. Draw the line of reflection.

   a.

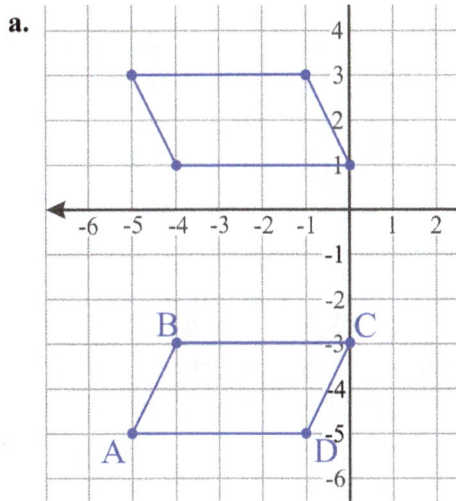

   b.

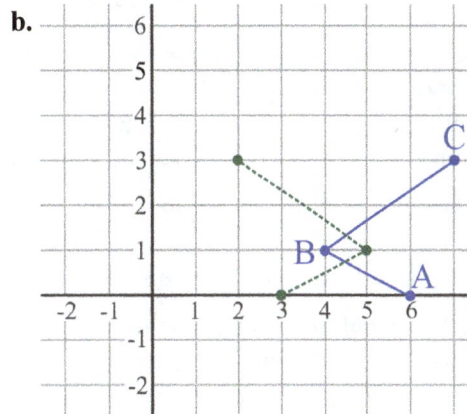

8. Draw the vertical line across which point D will be reflected onto D', and reflect the L-shape figure across that line.

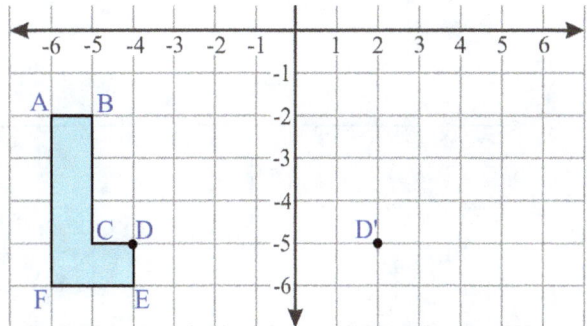

A quadrilateral was reflected in a horizontal line in such a manner that the coordinates of its vertices changed as follows:

   A $(-30, -25)$ → A' $(-30, -15)$

   B $(-28, -21)$ → B' $(-28, -19)$

   C $(-25, -21)$ → C' $(-25, -19)$

   D $(-23, -25)$ → D' $(-23, -15)$

Where is the line of reflection?

Translations and Reflections

In this lesson, we will explore sequences of transformations: that is, where several transformations are applied one after another.

For example, the dog on the right is first translated five units down, and then reflected in the green vertical line.

We call the line **the vertical line $x = -1$**, because the x-coordinate of all the points on the line is -1. (Note in particular that it crosses the x-axis when x is -1.) The equation $x = -1$ is actually the equation of that line.

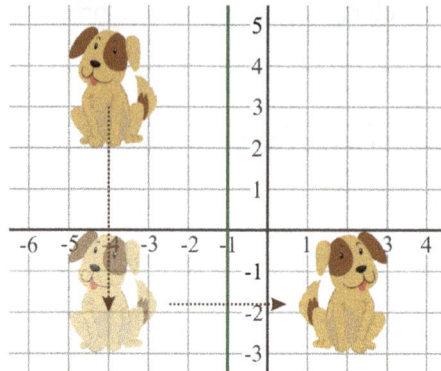

1. Does it matter in which order you do two separate transformations?

   a. First reflect the triangle across line n (the vertical line $x = 1$), then translate it five units up and one unit to the left.

   b. Now, first first translate the triangle five units up and one unit to the left. Then, reflect it across line n.

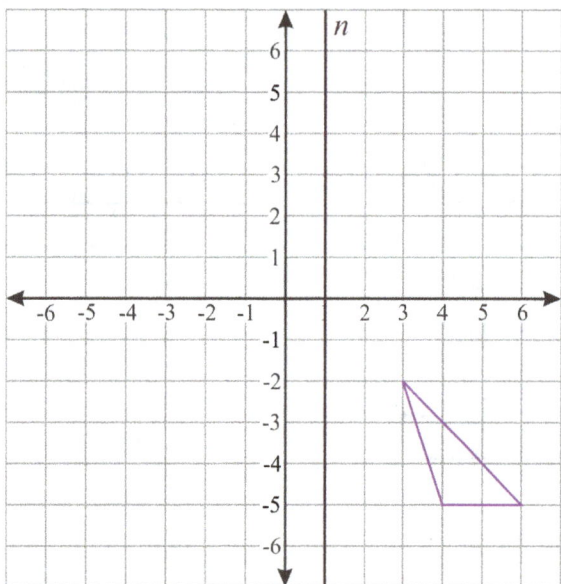

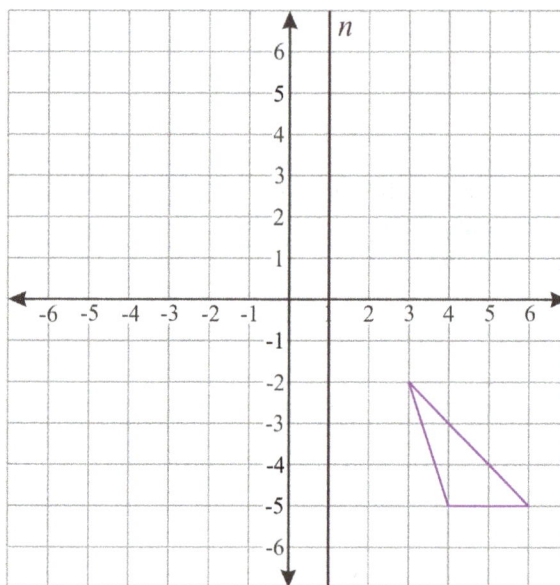

2. **a.** Reflect this figure twice: first in line m, then in line n.

   b. What single transformation would have produced the same result?

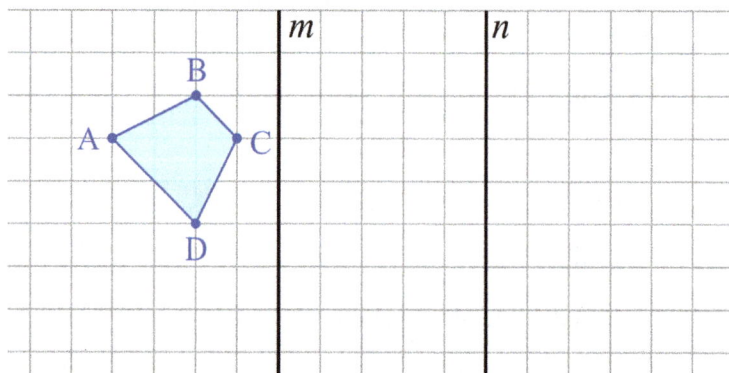

3. Describe a sequence of transformations that can map figure 1 to figure 2 using...

   a. a single reflection followed by a single translation.

   b. a single translation followed by a single reflection.

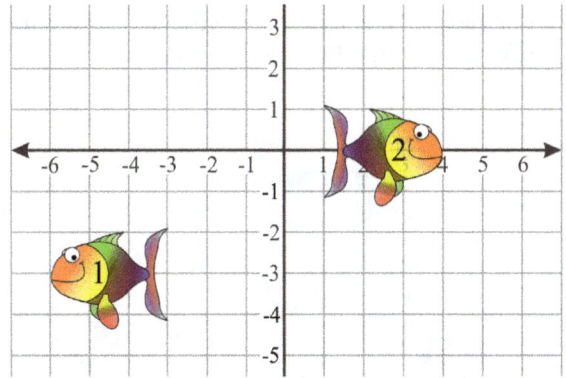

4. Describe a sequence of transformations that can map figure 1 to figure 2. Compare your answer to those of your friends or classmates, or find another solution yourself.

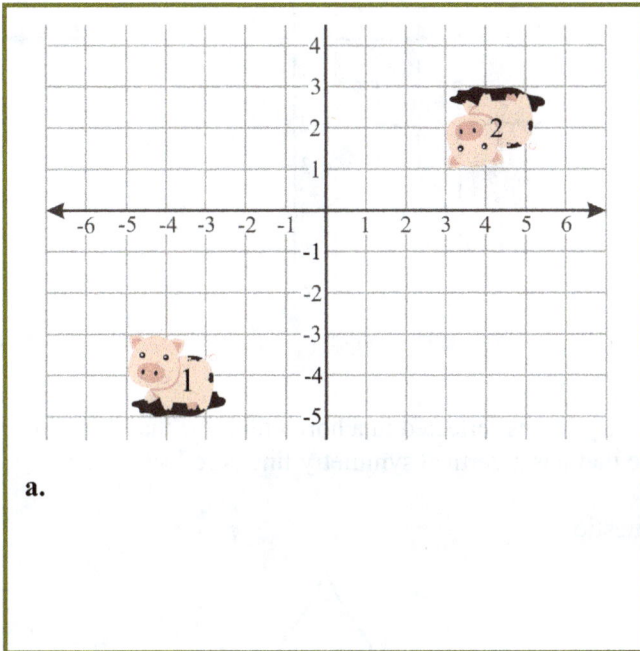

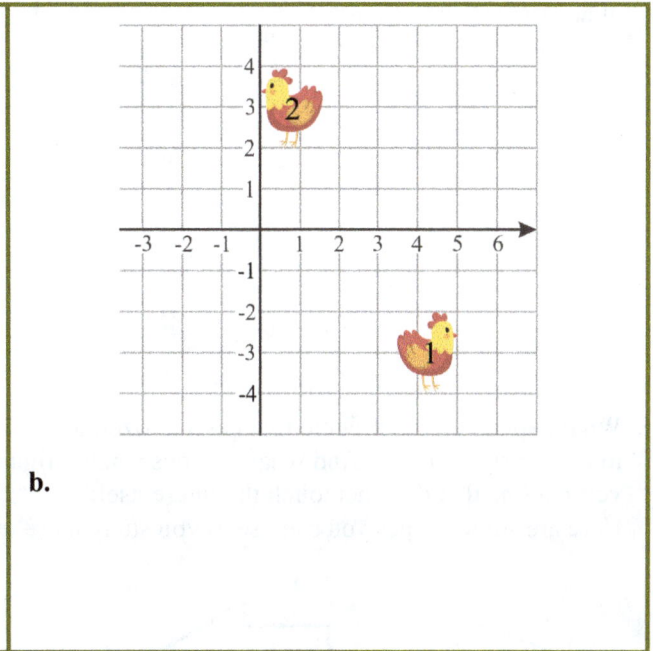

a.

b.

5. A triangle is first translated five units to the left, and then reflected in the horizontal line $y = 2$ (the horizontal line that crosses the y-axis at $y = 2$ and which contains all the points whose y-coordinate is 2).

   The vertices of the final figure are $(-5, 6)$, $(-4, 4)$, and $(-1, 5)$.

   Give the coordinates of the vertices of the original triangle.

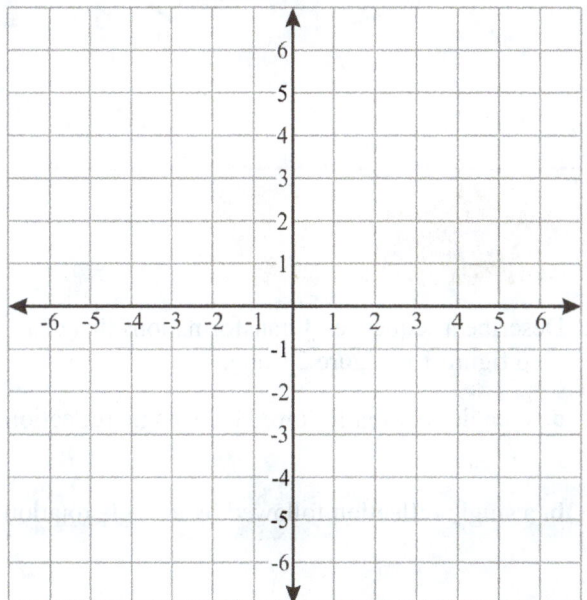

65

6. For each image, determine whether it is possible to obtain Figure 2 from Figure 1 using a *single* reflection. If yes, draw the line of reflection.

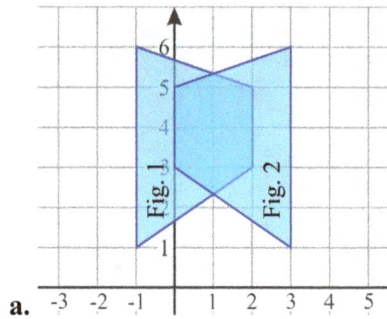

a.

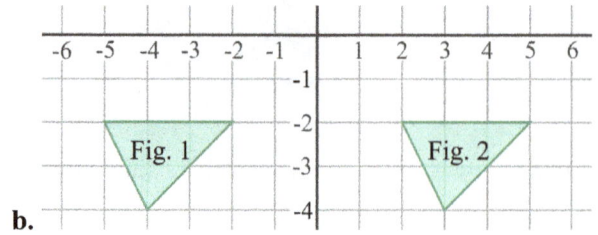

b.

7. Prove that figure 1 is congruent to figure 2 by explaining a sequence of transformations that maps figure 1 onto figure 2.

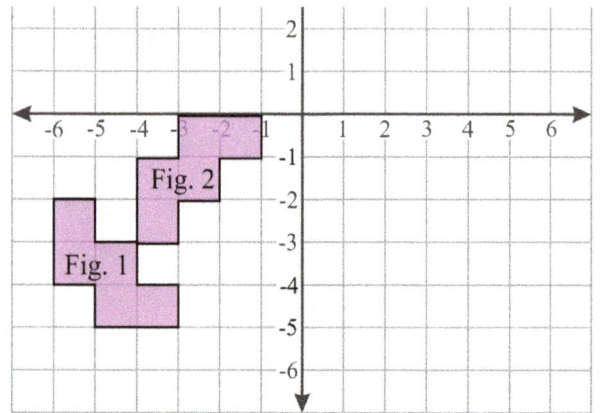

8. What happens when a figure that has a horizontal symmetry line is reflected in a horizontal line that does not touch the figure itself? And what happens when a figure that has a vertical symmetry line is reflected in a vertical line that does not touch the figure itself?
Here are some shapes you can use as you study these questions.

E   H   T

Puzzle Corner

Describe a sequence of transformations that can map figure 1 to figure 2 using...

a. a single rotation followed by a single reflection.

b. a single reflection followed by a single rotation.

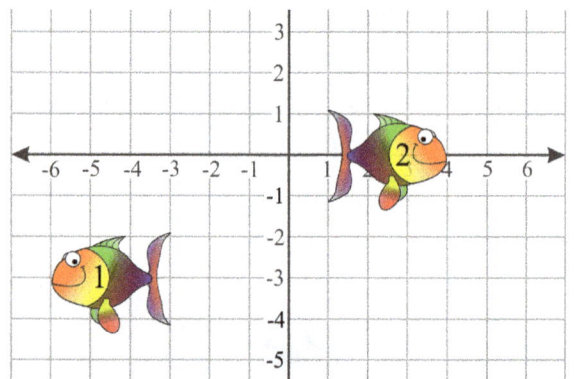

Rotations in the Coordinate Grid

Remember? In a **rotation**, a point moves along a circular arc around another point called **the centre of rotation**.

Rotations are measured in degrees. In this picture, P is rotated 125 degrees counterclockwise around point X, to become P'.

This means that the angle PXP' = 125°.

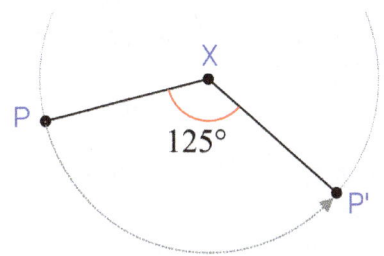

Something special happens to the coordinates of a point when it is rotated 90° *around the origin*, or point (0,0).

On the right, points A, B, and C are rotated 90° around the origin. Can you see how their coordinates change?

$$A(-5, 5) \rightarrow A'(5, 5)$$

$$B(-2, 5) \rightarrow B'(5, 2)$$

$$C(-5, 4) \rightarrow C'(4, 5)$$

Each time, the x and y-coordinates switch and adopt new signs that reflect the new quadrant they are in.

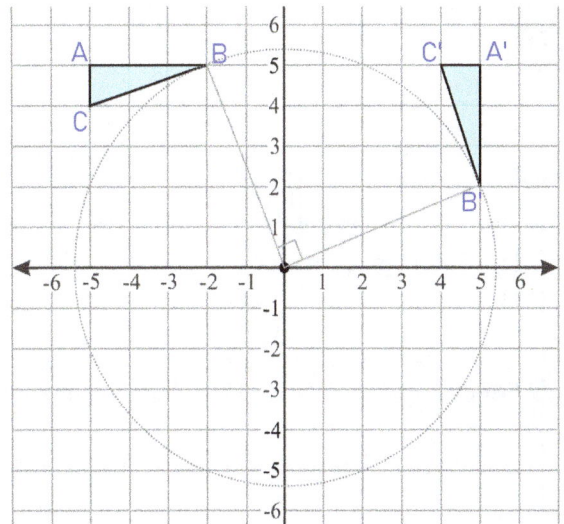

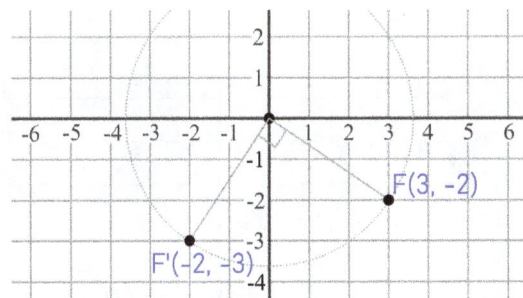

Example 1. When point F(3, −2) is rotated 90 degrees clockwise around the origin, the 3 and 2 are switched.

To know whether the new coordinates are positive or negative, look in which quadrant they are. Since F' is in the third quadrant where both coordinates are negative, the coordinates of F' are (−2, −3).

1. **a.** Rotate triangle A'B'C' clockwise another 90 degrees around the origin, to become triangle A"B"C".

   b. Now do it one more time: rotate triangle A"B"C" clockwise 90° around the origin.

   To help you, you can draw a line from the centre of rotation to the point you are rotating, and another line from the centre of rotation to the image point.

   These two lines created as if a "flag" in the angle of rotation (in this case, 90°).

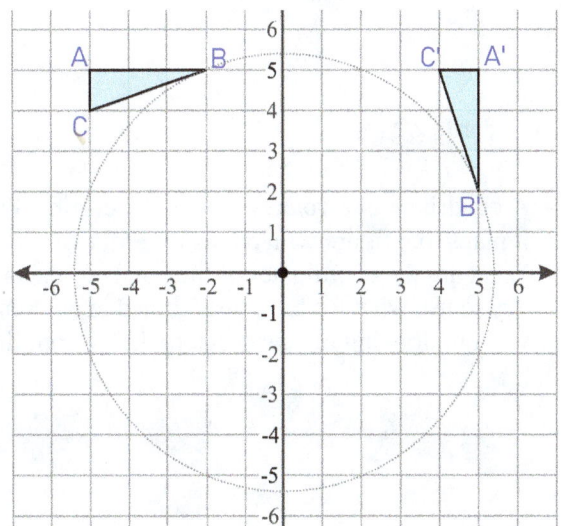

2. **a.** Rotate kite ABCD counterclockwise 90 degrees around the origin, to get its image, kite A'B'C'D'.

   b. Now rotate A'B'C'D' counterclockwise 90 degrees around the origin. Call the resulting figure A"B"C"D".

   c. Kite A"B"C"D" is the image of kite ABCD under

   a _____-degree rotation around the origin.

   Now compare the coordinates of kite ABCD and kite A"B"C"D". What do you notice?

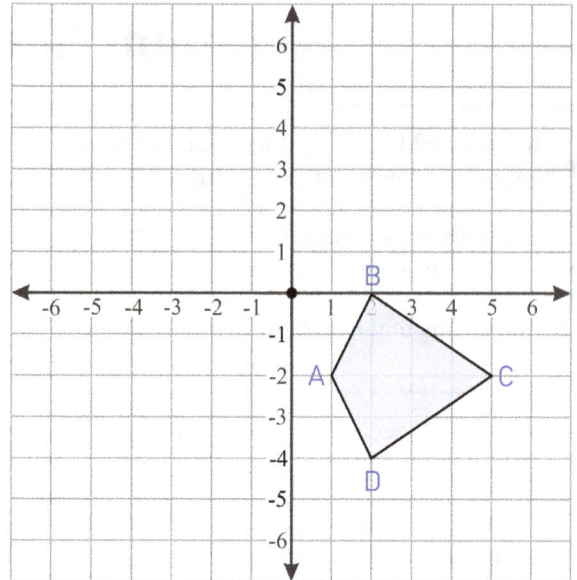

Based on your work in 2(c), fill in:

> In a 180-degree rotation around the origin, a point P is mapped onto point P' with coordinates that are _____ of the coordinates of P.
>
> For example, point $(-2, 1)$ is mapped onto point $(2, -1)$.

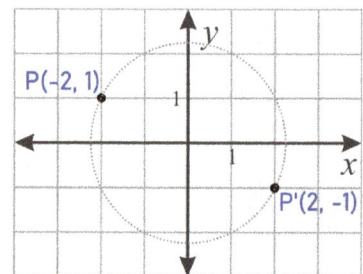

3. Angle GHI is rotated 90° clockwise around the origin.

   a. Which of the figures 1, 2, or 3 is an image of angle GHI under such rotation?

   b. Draw the image of ∠GHI under a 180-degree rotation around the origin.

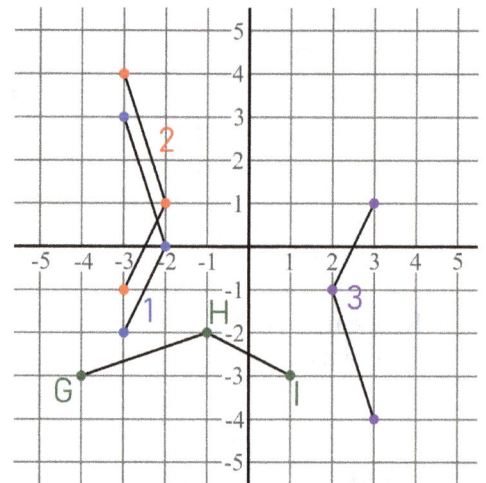

4. A quadrilateral is rotated around the origin 180 degrees so that it maps to a figure with points $(5, -6)$, $(3, -2)$, $(2, 0)$, and $(0, -4)$. What are the coordinates of the points of the original quadrilateral? Try to use what we have just learned to answer the question, without drawing the quadrilateral in the coordinate grid.

We can also rotate a point or a figure around a point other than the origin.

Here, point A is rotated around X clockwise 90 degrees. Notice that the distance from A to X and from A' to X are the same. In other words, AX and AX' have the same length. (Why?)

This fact makes it easy to find the coordinates of the rotated point *if* the rotation is 90, 180, or 270 degrees, and *if* the distance in question is exactly so many units along the gridlines.

In the other example, B is rotated around Y 90° counterclockwise. The distance of 2 units stays the same. We can see the coordinates of B' are (1, −4).

To calculate the final coordinates of a point when the angle of rotation is different than 90, 180, or 270 degrees is beyond the scope of this course. But what you learn here will enable you to rotate not only points, but also simple figures in the coordinate grid.

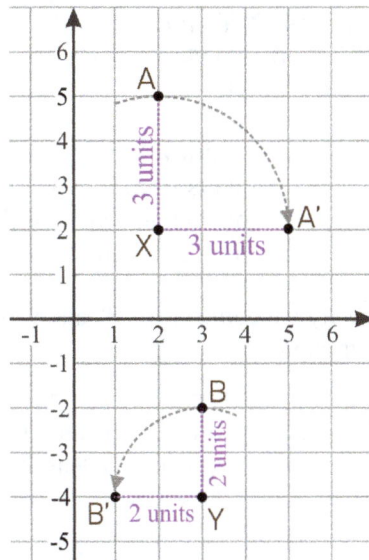

5. Draw the images of the given points under the following rotations:

   a. point A, around point X, 90° counterclockwise

   b. point B, around point X, 90° clockwise

   c. point C, around point Y, 90° clockwise

   d. point D, around point Y, 180° counterclockwise.

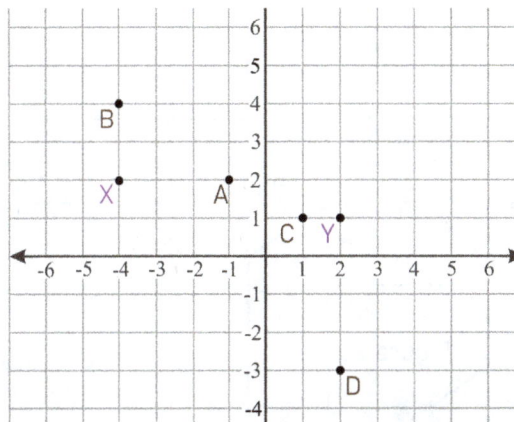

6. **a.** Points M and O are rotated 90 degrees clockwise about point Z. Draw M' and O', their corresponding image points.

   b. Connect M, N, and O to form a triangle.

   c. Figure out the location of N' under the same rotation, based on the location of M' and O'.

   d. Connect M', N', and O' to form a triangle.

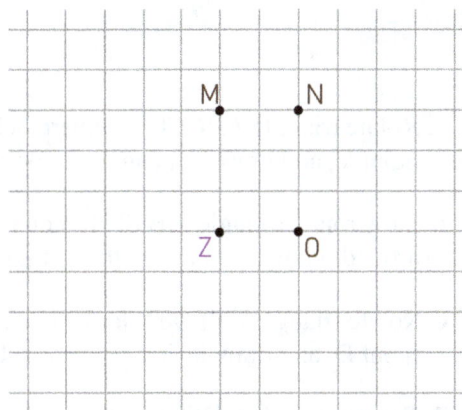

Puzzle Corner

Explain what single transformation is needed to map Figure 1 onto Figure 2. Be specific in your explanation.

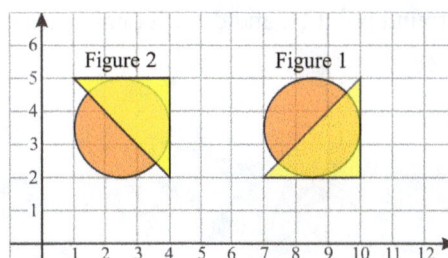

Figure 2   Figure 1

Example 2. To rotate triangle ABC around C (which is one of its vertices) 90° clockwise, we first rotate point B around C. It maps to B'(2, −3).

Then, we can draw A' at (2, −1) based on the facts that A' will be two units from B', just like A is two units from B, and also, we know that in a 90° rotation, $\overline{AB}$ being a horizontal line segment will map to a *vertical* line segment.

Lastly we connect A', B', and C' with line segments to get a triangle.

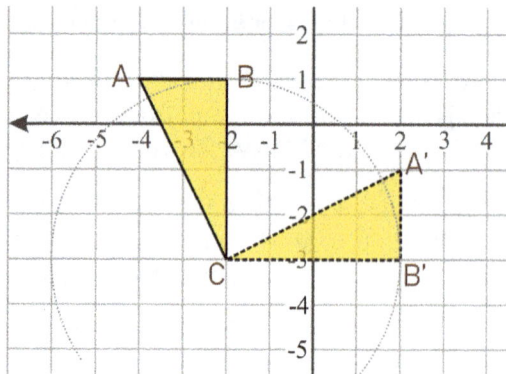

A note on notation: If P and Q are points, then $\overline{PQ}$ denotes a line segment between them, and PQ (without the top line) signifies the length of that line segment.

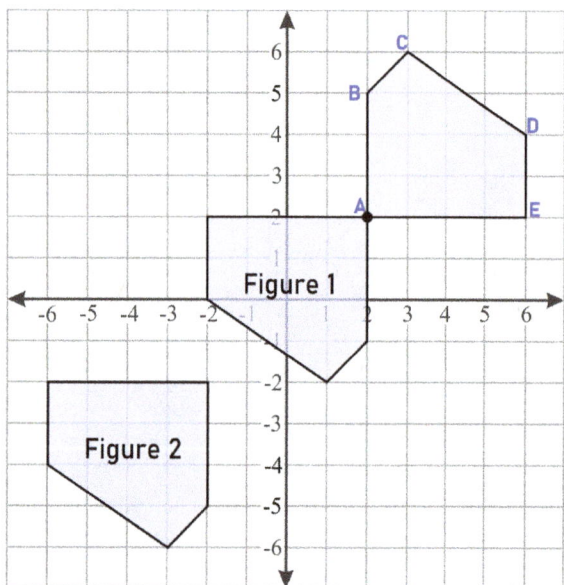

7. **a.** Describe the transformation that can be used to map pentagon ABCDE onto Figure 1. Give a detailed answer, not just the type of transformation.

b. Describe the transformation that can be used to map pentagon ABCDE onto Figure 2. Give a detailed answer, not just the type of transformation.

8. **a.** Rotate triangle ABC 90° counterclockwise around point C, and draw its image triangle A'B'C'.

b. Now rotate triangle A'B'C' 90° counterclockwise around point C', and draw the triangle A"B"C".

c. Rotate triangle DEF 90° clockwise around point D, and draw its image triangle D'E'F'.

d. Rotate triangle GHI 90° counterclockwise around point H, and draw its image triangle G'H'I'.

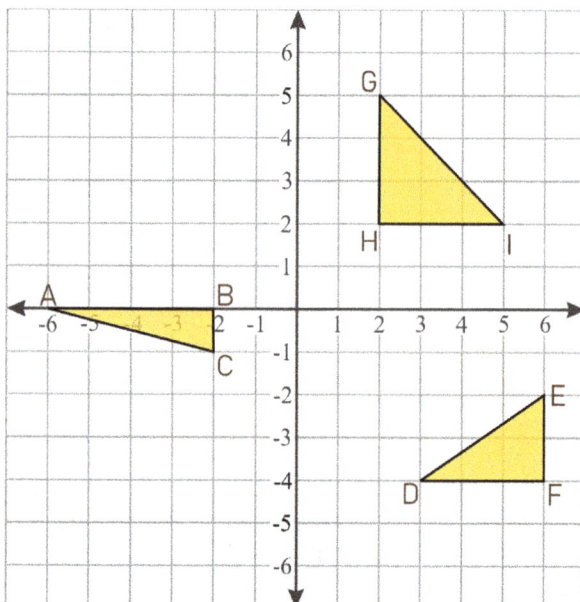

Sequences of Transformations

Here, the L-shape is first reflected across the y-axis. Then the resulting figure is rotated around point B' 90 degrees counterclockwise.

This is a **sequence** or a **composition** of two transformations: first a reflection, then a rotation.

The final figure is congruent to the original. Why?

It is because each individual transformation preserves congruence, so a sequence of them does also.

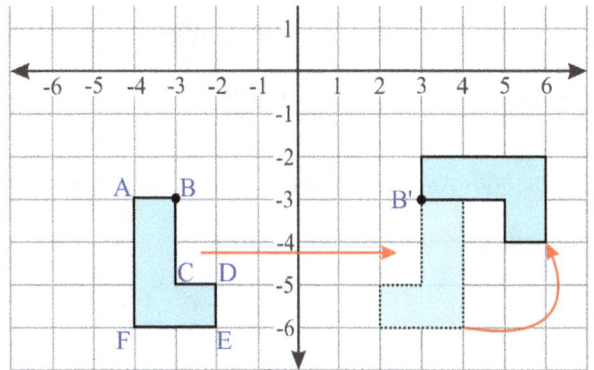

Note 1: The transformations we will use in this lesson include translations, reflections in vertical or horizontal lines, and rotations of 90, 180, or 270 degrees around the origin or around a vertex of the figure.

Note 2: Different sequences of transformations will often produce the same end result.

1. **a.** Describe a sequence of two transformations that can map figure 1 to figure 2.

   b. Can you find another sequence?

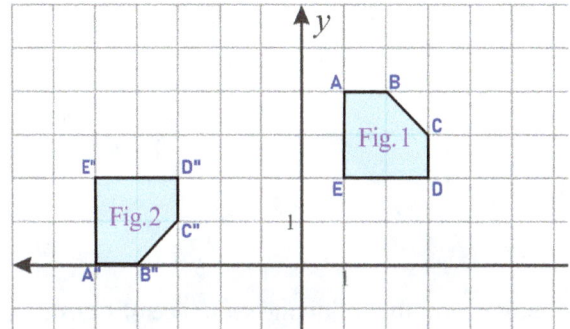

2. **a.** Describe a sequence of transformations that can map parallelogram ABCD onto figure 1.

   b. Now describe another, different sequence of transformations that can map parallelogram ABCD onto figure 1.

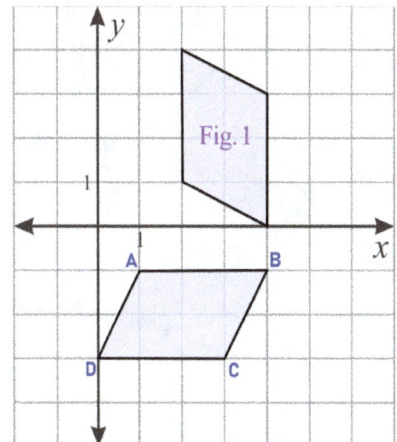

3. Andy wrote the following proof to prove that figures 1 and 2 are congruent. It has two errors though. Correct the errors in his proof.

A rotation 90 degrees clockwise around point E,

followed by a translation five units down and

three to the left transforms Figure 1 to Figure 2.

Since both rotations and translations preserve

congruence, the two figures are congruent.

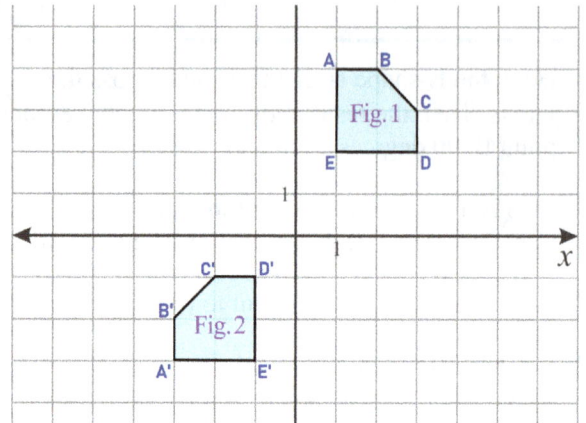

4. Write a proof to show that triangle ABC is congruent to Figure 2.

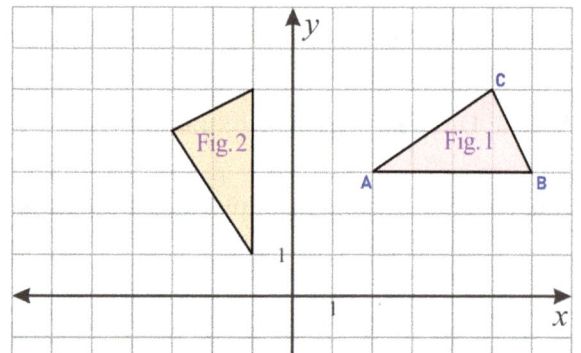

5. **a.** Perform the following sequence of transformations to triangle ABC:

First reflect it in the horizontal line $y = 1$. Then rotate it around the origin 180 degrees. Then translate it four units up and two to the left.

   b. Find a sequence of *two* transformations that does the same as the sequence in (a).

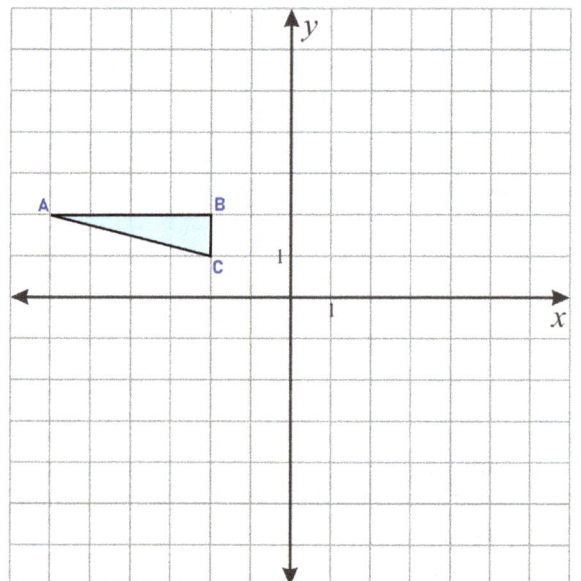

6. Draw any *right* triangle in the highlighted quadrant of the coordinate grid (the third quadrant). Call it triangle Z.

   a. Reflect it in the *y*-axis, to become triangle Z'. Then reflect triangle Z' in the *x*-axis, to become triangle Z".

   b. Will you get the same final result if you first reflect triangle Z in the *x*-axis, then reflect its image in the *y*-axis?

   c. What single transformation between triangle Z and triangle Z" would have produced the same result?

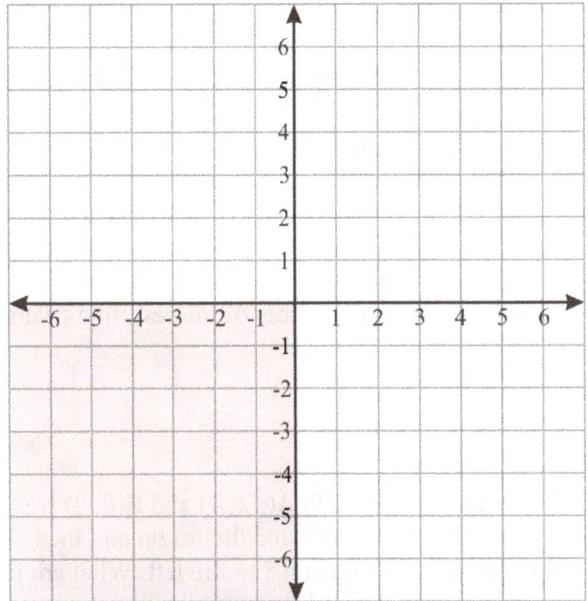

Puzzle Corner

During the night, the stars and the constellations seem to rotate around a certain point, which is very near the North Star. If you leave the camera's shutter open for a long time (hours) and take a picture of the stars, you will see *star trails* — the paths these stars travel along as they rotate around the North Star.

Below, use transparent paper to rotate the Big Dipper around the North Star in such a manner that Dubhe (one of the stars) maps to the point marked with (Dubhe).

Afterwards, if you'd like, you can also draw the star trails using a compass.

(Dubhe)

North Star

Dubhe

73

Sequences of Transformations, Part 2

Note: You can use the grid to help you with the following problems, but try to solve them without using it.

1. A triangle with vertices A(1, 2), B(5, 3), and C(4, 1) was first reflected in the y-axis and then translated 6 units down and two to the right. What are the coordinates of the vertices of the resulting triangle?

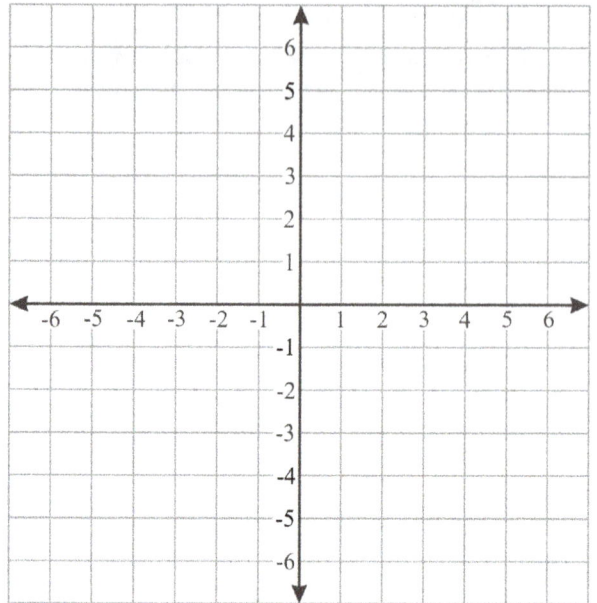

2. Line segment $\overline{AB}$ with A(−2, 4) and B(0, 2) was rotated 180 degrees around the origin and then translated 7 units up and 5 to the left. What are the coordinates of the end points of the line segment after these transformations?

3. A quadrilateral was first rotated around the origin counterclockwise 90 degrees, and then reflected in the x-axis. Its vertices are now at points (3, 5), (5, 2), (3, 1), and (2, 2). What were the coordinates of its vertices before these transformations?

4. Triangle ABC is as shown on the right. It will be rotated around the origin counterclockwise 90 degrees, then translated 5 units down, and lastly, rotated once again around the origin counterclockwise 90 degrees.

   Ashley claims that the transformed triangle's vertices are at (5, −2), (1, −3), and (2, −1). Is she correct? Explain.

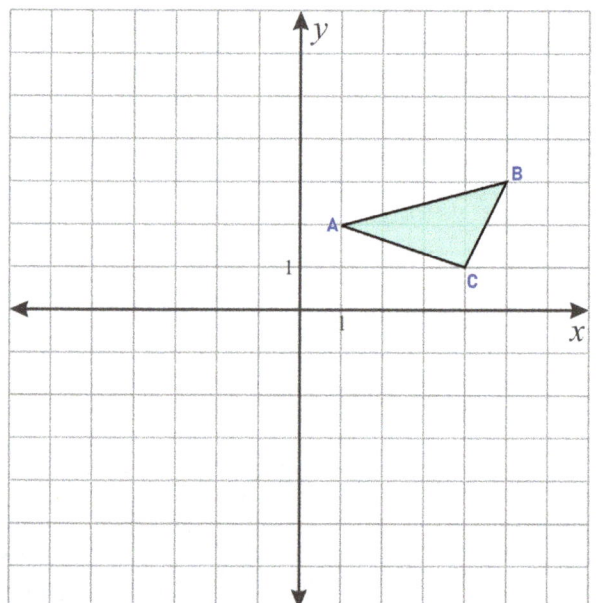

5. Greg says that the two rectangles are congruent because you can reflect rectangle ABCD in the *y*-axis and then move it five units up to map it onto rectangle A'B'C'D'.

Jenny says that's too complicated; you can simply translate rectangle ABCD five units up and two units to the left, and that does the job.

Who is correct, or are both correct? Why?

(Hint: Note the vertices carefully.)

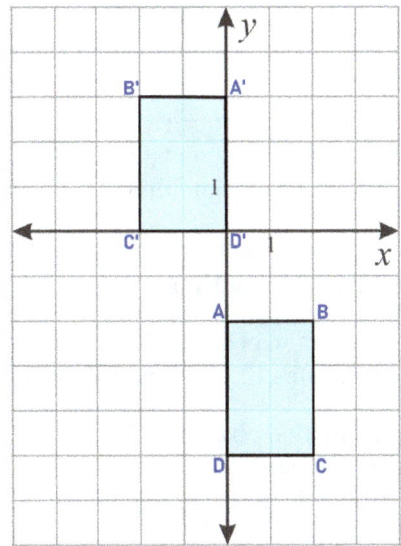

6. A quadrilateral with vertices H(−5, 2), I(−4, 4), J(−2, 4), and K(−4, 1) is reflected in the horizontal line *y* = 1, and then rotated around the origin 180 degrees. Find the coordinates of the transformed figure.

7. Triangle PQR underwent a translation, then a reflection. Study the coordinates to find out the details about each transformation, then fill in the missing coordinates.

Original figure	Translation	Reflection
P(−5, −2)	P'(−6, 3)	P''(4, 3)
Q(−3, −2)	Q'(−4, 3)	Q''(2, 3)
R(−4, 1)	R'(____ , ____)	R''(____ , ____)

8. Which of the figures 1, 2, 3, or 4 is the image of triangle DEF when it undergoes the following sequence of transformations?

1. Rotation 90° clockwise around D;

2. Rotation 180° around the origin;

3. Reflection in the *x*-axis;

4. Translation two units to the right.

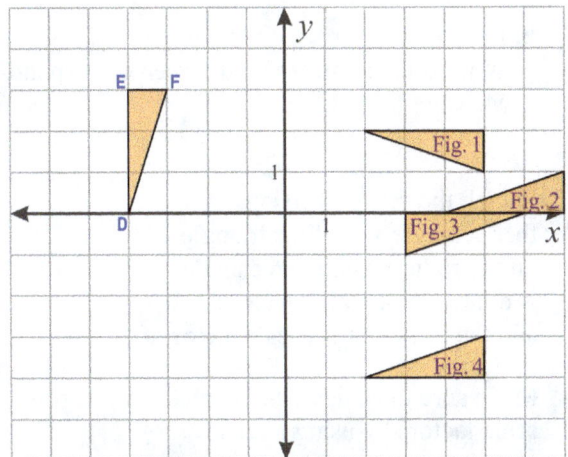

Dilations

A **dilation** is a transformation where the distances between points in a figure are multiplied by a certain factor, called **scale factor**.

For example, here the distances between the vertices (the side lengths) get multiplied by a factor of 1.5:

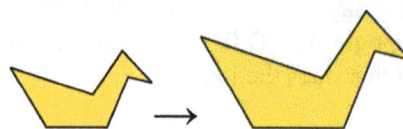

A dilation changes the size of a figure proportionally, maintaining its shape but not its size. If the scale factor is less than 1, the figure shrinks, and if it is more than 1, the figure is enlarged. (What about if it is 1?)

A dilation obviously does *not* preserve distances. It does, however, preserve the overall shape of the figure, and in particular, its angles. The dilated figure looks like a perfect enlargement or reduction of the original, as you will see in this lesson.

1. Which figures are dilations (proportional enlargements or reductions) of figure 1?

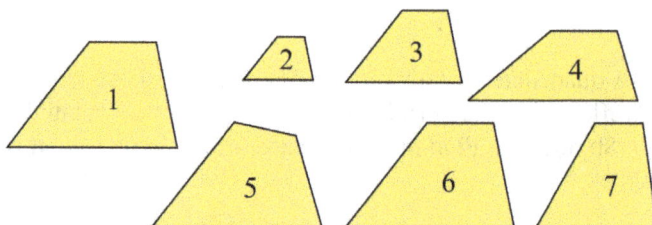

2. Triangle ABC is dilated to produce triangle DEF.

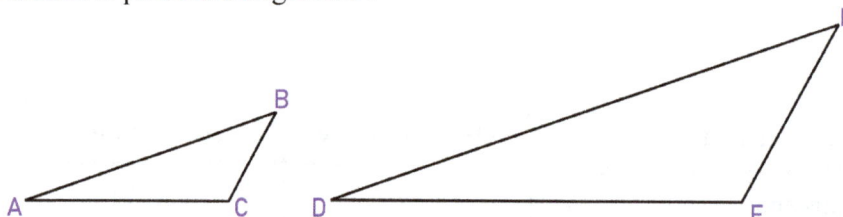

   a. Indicate the corresponding sides (list which side of △ABC corresponds to which side of △DEF).

   What is the relationship between corresponding side lengths? Measure to find out.

   b. Indicate the corresponding angles.
   What is the relationship between corresponding angles? Measure to find out.

3. Sketch the dilated versions of the rectangle and of the triangle on the right, when the rectangle is dilated by a scale factor of 1/2, and the triangle by a scale factor of 3.

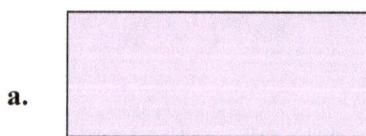

   a.

   Make sure to achieve the correct scale factors by using a ruler.

   b.

A dilation always has a fixed point, called **the centre of dilation**. Here, it is point O.

The centre point can be anywhere. When the scale factor is less than 1, the reduced image will be closer to O than the original. When SF is greater than 1, the dilated figure will not only be larger, but also be farther away from O. However, the dilation (the enlargement or shrinking) will work no matter where O is located.

Let's use scale factor 2 in this example to keep things simple. This means all distances will double. To find the image of point P under this dilation, follow these steps:

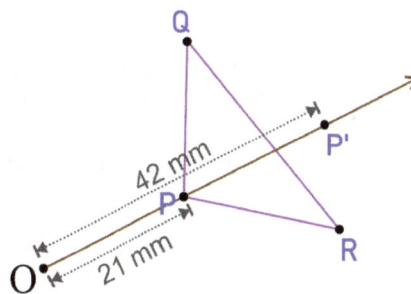

1. Draw **a ray** starting from O that goes through point P.
2. Measure the distance OP. In this example, it is 21 mm.
3. Multiply that distance by the scale factor, in this case 2, to get 42 mm.
4. Draw P' so that it is at the distance of 42 mm from O.

The diagram on the right shows how to draw the images of points Q and R using the same principle: Since the scale factor is 2, the distance from O to R' must be twice the distance between O and R. Similarly, the length of $\overline{OQ'}$ is twice the length of $\overline{OQ}$, and $\overline{OP'}$ is twice $\overline{OP}$.

Lastly, we join points P', R', and Q' with line segments to get a triangle that is an image of triangle PQR under this dilation.

Verify that the side lengths of triangle PQR indeed doubled!

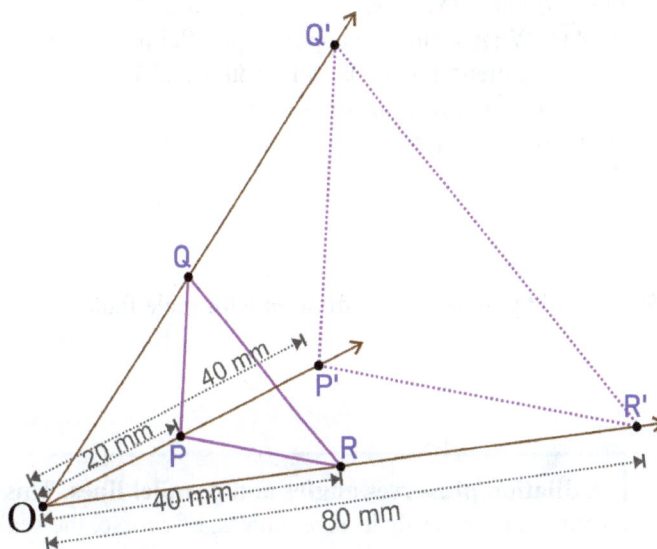

In summary, a dilation is described in terms of

- its centre point and
- the scale factor.

4. **a.** Draw the images of points F, G, and H under a dilation with centre point O and scale factor 3. Name them F', G', and H'.

   b. Join the points F, G, H and also their images to get two triangles. Measure to check whether the side lengths of triangle F'G'H' are three times those of triangle FGH or not. If you drew accurately, they should be!

   c. How does angle G'H'F' compare with angle GHF?

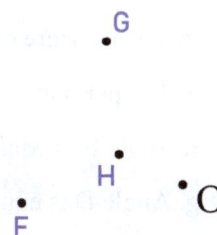

A dilation can also be drawn using one of the vertices of the figure as the centre of dilation. We say the figure is dilated *from* that point.

Here, trapezoid PQRS is dilated from point Q (in other words, Q is the centre of dilation), and with a scale factor of 2.5

Notice that the image of point Q under this dilation is Q itself!

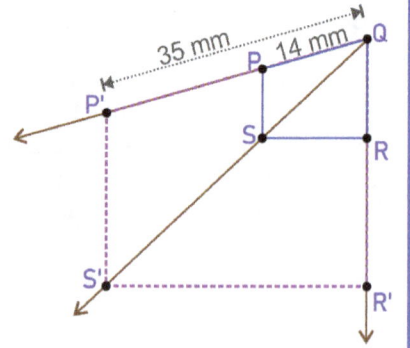

5. **a.** Draw a dilation of this rectangle from point B and with a scale factor of 3/4.

b. Figure ABCD is a rectangle, which means the sides $\overline{AB}$ and $\overline{CD}$ are parallel, and so are $\overline{BC}$ and $\overline{AD}$. What seems to happen to parallel lines (or line segments) in a dilation? You can also check the example at the top of this page, where $\overline{QR}$ and $\overline{PS}$ are parallel sides of the trapezoid PQRS.

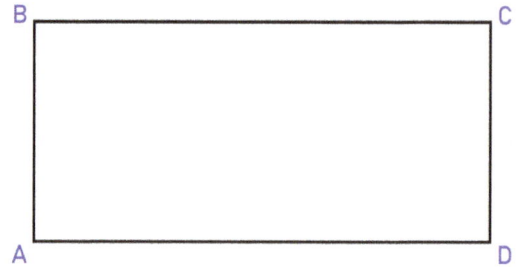

6. How would you describe a dilation with scale factor 1?

A dilation preserves angles and parallel lines. This means it preserves the shape of a figure (but not necessarily its size, unless, of course, the scale factor is 1).

A dilation of a figure always produces a figure that is **similar** to the original. In practical terms, two figures being similar means they have the same shape, but are not necessarily of the same size.

7. Figure A'B'C'D'E' is a dilation of figure ABCDE with scale factor 2/3. Angles A and E are right angles. Check all the statements that are true.

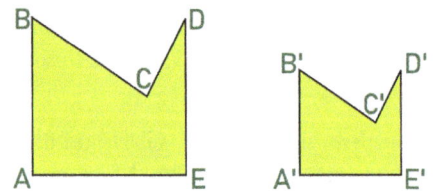

a. Angle B'A'E' is a right angle.

b. If AE = 1 inch, then A'E' = 1/2 inch.

c. $\overline{A'B'}$ is parallel to $\overline{D'E'}$.

d. The measure of ∠EDC is 2/3 of the measure of ∠E'D'C'.

e. The perimeter of ABCDE is 1.5 times the perimeter of A'B'C'D'E'.

f. Angle B is equal to angle B'.

g. Angle D is equal to angle B'.

Dilations in the Coordinate Grid

Example 1. An angle with vertex at (3, 2) is dilated with origin as centre and a scale factor of 2.

Recall that the coordinates of a point indicate the horizontal and vertical distance between the point and the origin (point 0, 0). Since in a dilation, all distances are multiplied by the scale factor, the horizontal and vertical distances are too. This makes it easy to find the coordinates of any point under this dilation: both the *x* and *y*-coordinate of each point are simply doubled.

Verify this fact for the other two points of the original figure.

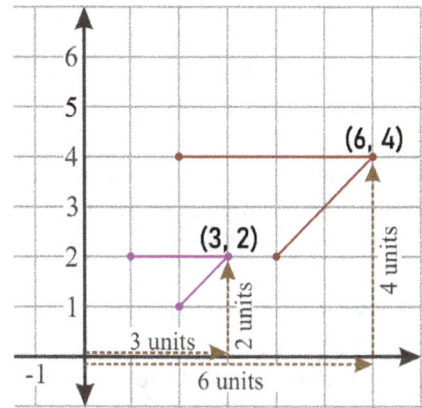

If the centre point of a dilation is <u>the origin</u>, we can simply **multiply the coordinates of a point by the scale factor** to get the coordinates of the image of the point.

1. In each case, dilate the figure with origin as centre and with the given scale factor.

a. scale factor 2

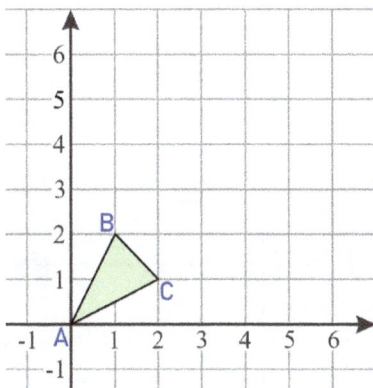

b. scale factor 1/3

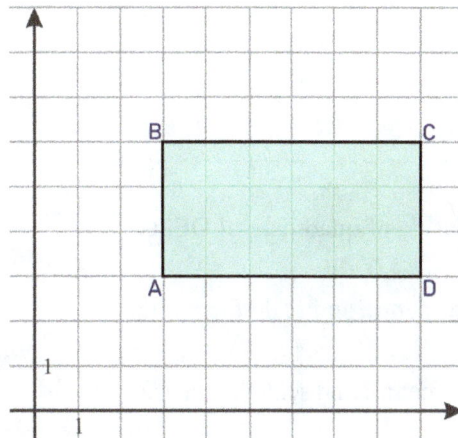

c. scale factor 1/2

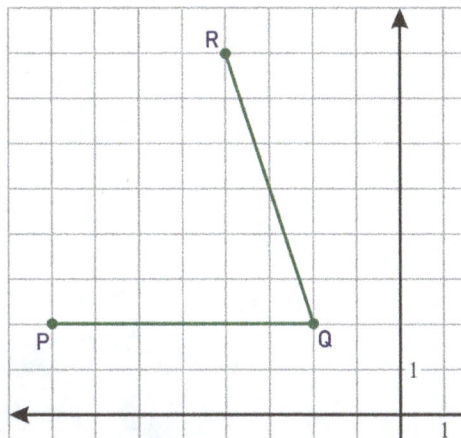

d. scale factor 3

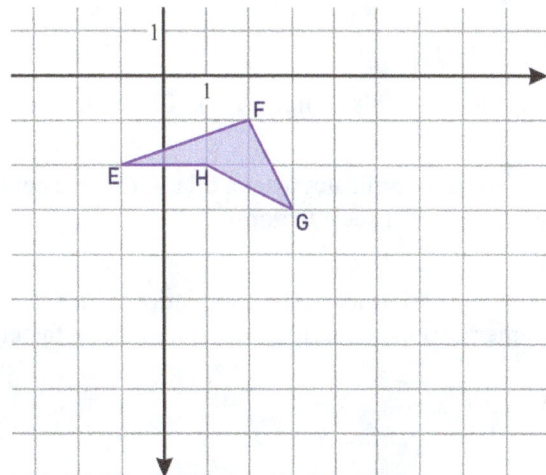

Example 2. If the centre of dilation is not the origin, the calculation of the coordinates is not as straightforward. But it is still fairly easy to draw the dilated figure, using the horizontal and vertical distances.

Here, triangle ABC is dilated from point B and with a scale factor of 2. The distance between B and C is 1 unit horizontally and two units vertically. This doubles so that the distance between B' (which is the same as B) and C' is two units horizontally and four vertically.

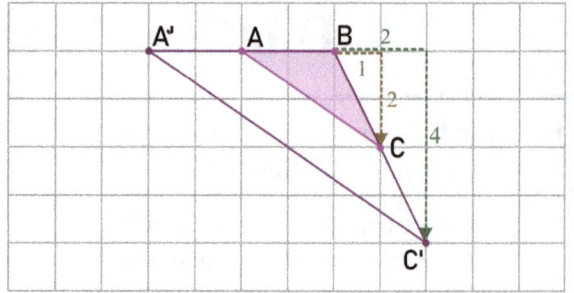

2. Draw a dilation of triangle ABC...

a. from point A and scale factor 2

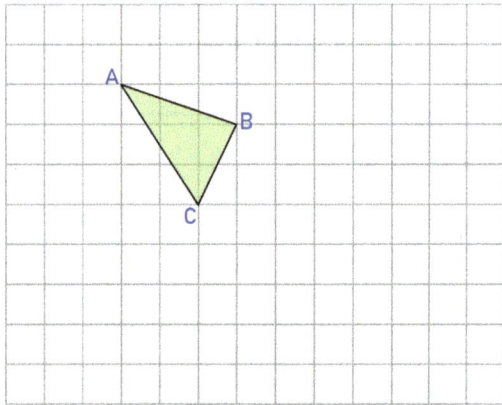

b. from point B and scale factor 3

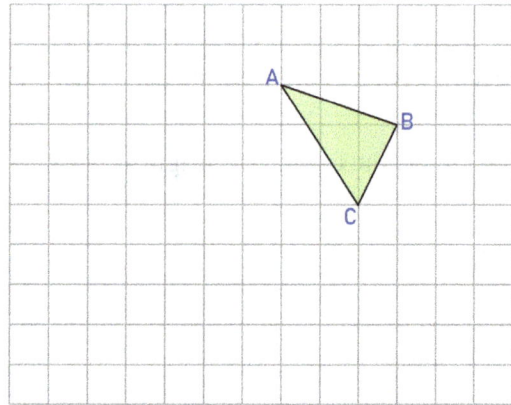

3. Draw a dilation of quadrilateral DEFG:

   a. with D as centre and scale factor 1/2;

   b. with F as centre and scale factor 1/2.

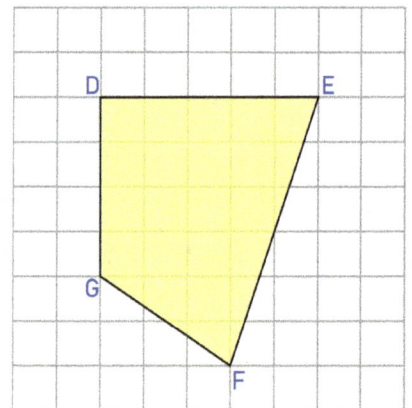

4. The coordinates of a kite are $(-3, 3)$, $(0, 6)$, $(9, 3)$, and $(0, 0)$.

   a. What are its coordinates after a dilation with a centre at the origin, and a scale factor of 2/3?

   b. What are its coordinates after a dilation with a centre at the intersection of the diagonals, and a scale factor of 2/3?

5. Describe each dilation in terms of its centre point and its scale factor.

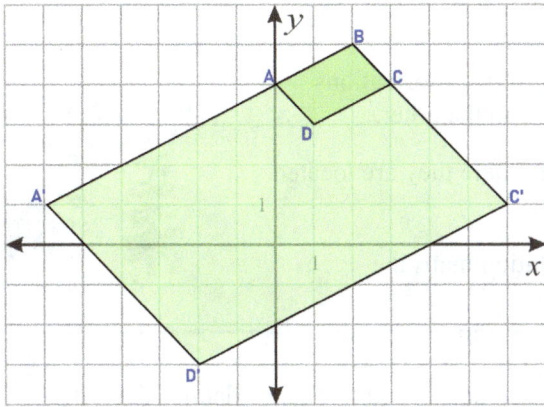

a. Centre point: _____   Scale factor: _____

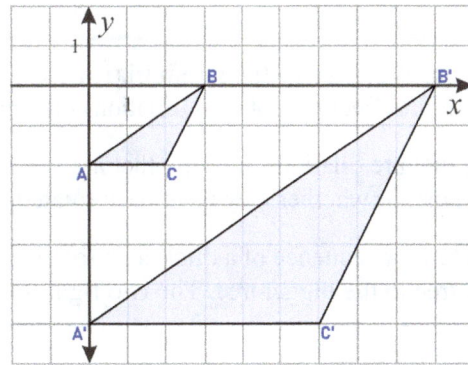

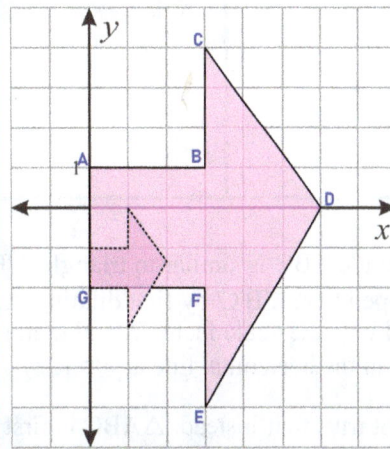

b. Centre point: _____   Scale factor: _____

6. Describe each dilation in terms of its centre point and its scale factor. This time, the figure becomes smaller.

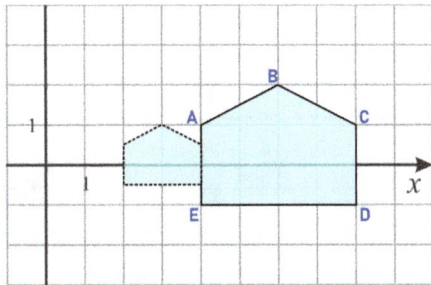

a. Centre point: _____   Scale factor: _____

b. Centre point: _____   Scale factor: _____

7. Both figure 1 and figure 2 are dilations of rectangle ABCD, with a scale factor of 2.

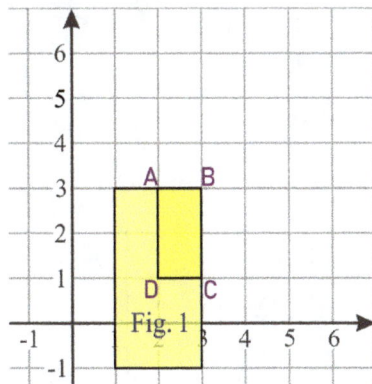

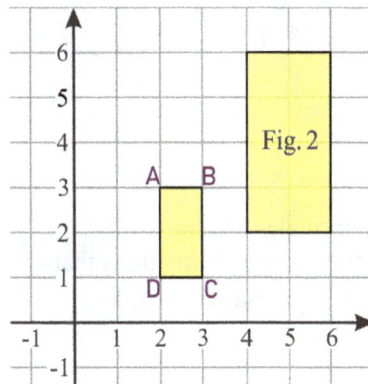

Explain why the dilated images are different from each other, even though the scale factor is the same.

Similar Figures, Part 1

Definition: We call two figures **similar** if there is a sequence of transformations (translation, reflection, rotation, dilation) that maps one figure to the other.

Figures that are dilations of each other are similar, no matter where they are located in the plane, or whether they have been rotated or reflected.

Example 1. A sequence of a dilation, a rotation, and a translation maps the smaller tree to the bigger tree. The two figures are similar.

1. State the transformations that can map figure 1 to figure 2. You don't need to include details about the transformations, such as the scale factor, the exact line of reflection, or the amount of translation or rotation.

2. Henry says that triangle ABC is similar to triangle A'B'C' because △ABC can be mapped to △A'B'C' by first dilating △ABC with origin as centre and with the scale factor 2, and then reflecting the resulting figure in the horizontal line at $y = 1$.

   Harry says that's not true, that instead, △ABC is first reflected in the x-axis, and then dilated with B' as centre point, with scale factor 2.

   Whose proof is correct, or are both correct?

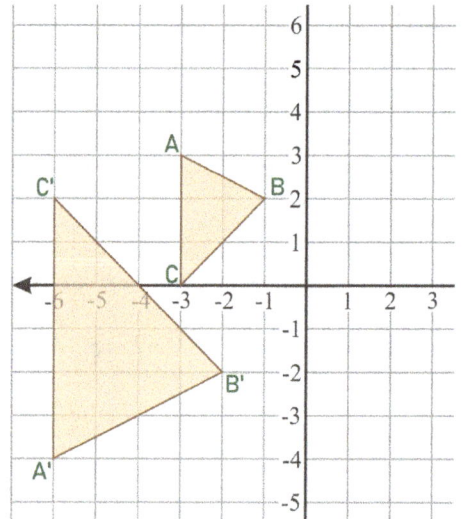

3. Show that the two triangles are similar by describing a sequence of transformations that could map △ABC to the smaller triangle.

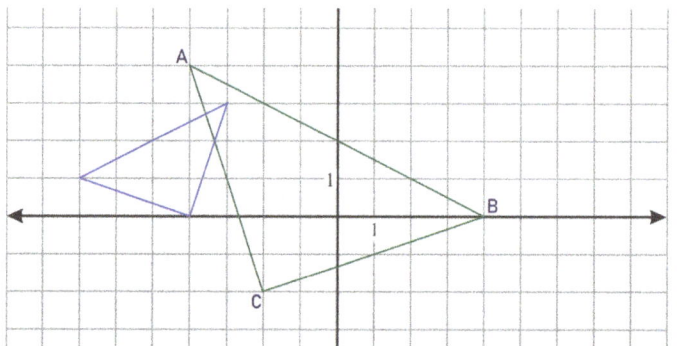

4. Parallelogram ABCD underwent the following transformations :

   1. A 90° rotation clockwise around the origin.

   2. Translation 3 units to the right and 4 units down.

   3. Dilation centred at B" with scale factor 1/2.

   What are the coordinates of the image of point D (D''') after all these transformations?

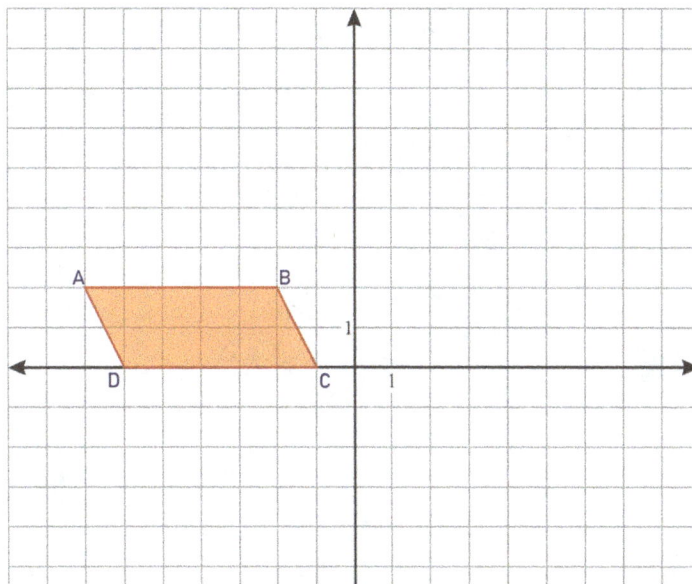

5. Triangle PQR underwent two transformations. Study the coordinates to find out the details about each transformation. Then describe each transformation in detail. Use grid paper if necessary.

Transformation 1:

Original figure	Transformation 1	Transformation 2
P(−4, 0)	P'(−1, 0)	P"(1, −4)
Q(0, 4)	Q'(0, 1)	Q"(2, −3)
R(−8, 4)	R'(−2, 1)	R"(0, −3)

Transformation 2:

6. **a.** If two figures are congruent, are they also similar? Explain your reasoning using the definition of similarity.

   b. Is it true that two similar figures must also be congruent? Why or why not?

7. Figure EFGH underwent a dilation, then a reflection. Study the coordinates to find out the details about each transformation, then fill in the missing coordinates. Use grid paper if necessary.

Original figure	Dilation	Reflection
E(−1, −1)	E'(−5, −2)	E"(−5, 2)
F(3, 0)	F'(3, 0)	F"(3, 0)
G(3, −2)	G'(3, −4)	G"(____ , ____)
H(2, −2)	H'(____ , ____)	H"(____ , ____)

8. Are the two figures similar? If yes, give a proof of that by giving a sequence of transformations (with details) that maps one to the other.

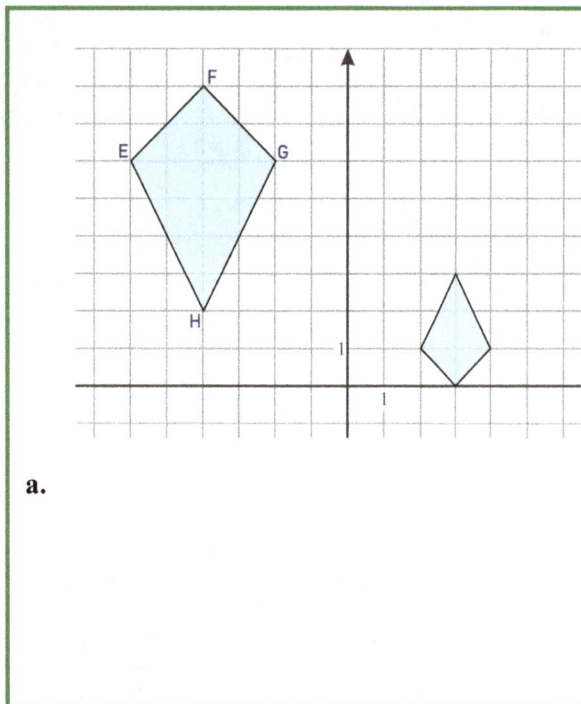

a.

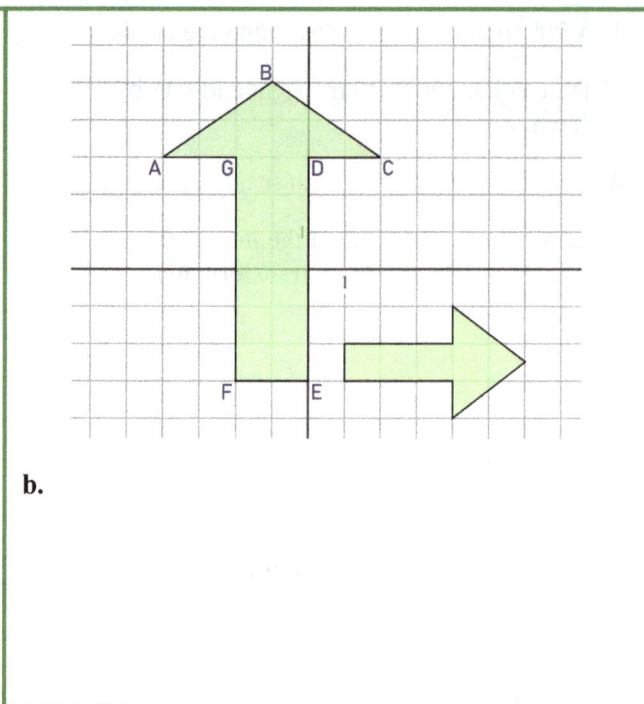

b.

9. Which statement is true?

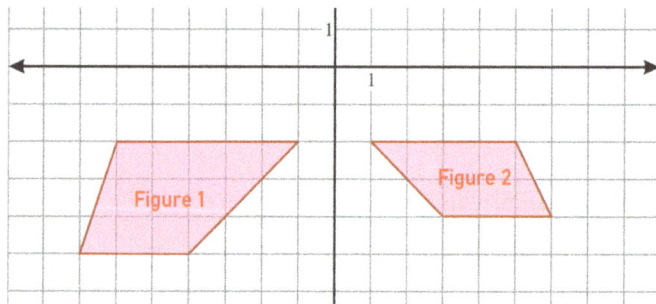

(1) A reflection in the *y*-axis, followed by a dilation, will transform Figure 1 to Figure 2, proving the two are similar.

(2) There is no sequence of transformations that will map Figure 1 to Figure 2, making the two figures neither congruent nor similar.

(3) A rotation 180 degrees about the origin, followed by a translation, followed by a dilation, will transform Figure 1 to Figure 2, proving the two are similar.

(4) A translation, then a reflection will make Figure 1 map to Figure 2, proving they are congruent.

Puzzle Corner Figure WXYZ underwent two mystery transformations. Study the coordinates to find out the details about each transformation, then fill in the missing coordinates. Grid paper can help.

Original figure	Transformation 1	Transformation 2
W(−6, 0)	W'(2, 0)	W"(2, 0)
X(−3, 1)	X'(−1, 1)	X"(3, 3)
Y(−3, −3)	Y'(−1, −3)	Y"(−1 , 3)
Z(−6, −2)	Z'(____ , ____)	Z"(____ , ____)

Similar Figures, Part 2

Recall that a dilation preserves angles, and the distances between points are multiplied by the scale factor of the dilation. This means that when two figures are similar...

- their corresponding angles are equal.

- their corresponding sides are **proportional**; in other words **in the same ratio**. In terms of the image, this means that $a/A = b/B = c/C$. This ratio is called the **scale ratio** or the **similarity ratio**.

- Proportionality also means that $a:b = A:B$, $a:c = A:C$, and so on.

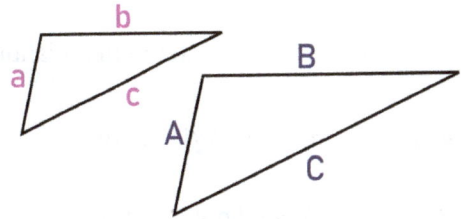

Example 1. Trapezoid ABCD is dilated by scale factor 1.25.

   a. What is the scale *ratio*?

   b. If A'D' = 25 cm, what is AD?

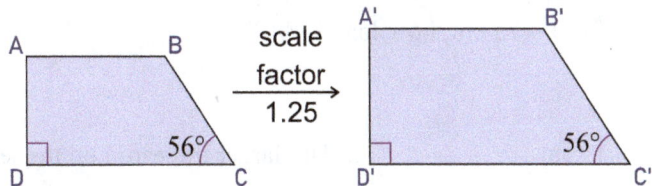

First, we write the scale factor as a fraction, in lowest terms: 1.25 equals 5/4 as a fraction. Now we switch the 5 and 4 and write it as a ratio: the scale *ratio* is 4:5.

The reason for the switch is that in the scale ratio, we want the *first* member of the ratio (4) to correspond to the original figure, which in this case is smaller, and the second (5) to the scaled, bigger figure.

The corresponding sides of the two figures are in the ratio 4:5. This means that AD:A'D' = 4:5, and AB:A'B' = 4:5, and also every other ratio of corresponding sides equals 4:5. Using fractions, we can also express this as AD/A'D' = 4/5 and similarly for the other corresponding sides.

To find AD, we either multiply or divide by the scale factor. Since we are going from the bigger figure to the smaller, and our scale factor is 1.25, or 5/4, we will divide A'D' by it: 25 cm / 1.25 = 20 cm.

Or, you could use fraction division: $25 \text{ cm} \div \dfrac{5}{4} = 25 \text{ cm} \cdot \dfrac{4}{5} = 100 \text{ cm} / 5 = 20 \text{ cm}$.

Generally speaking, when the scale ratio is $a:b$, to go from one shape to the other we either multiply by a/b, or by b/a, depending which direction we are going in.

Yet another way to solve for the side lengths is to use a proportion (see example 2).

Example 2. Parallelogram ABCD is similar to parallelogram EFCG. Write a proportion to solve for EF.

Look carefully how we set up the proportion. We have marked the unknown side with x. The shorter sides of both parallelograms are in the numerators and the longer ones in the denominators. You could have them vice versa, if you'd like. Both ways will work.

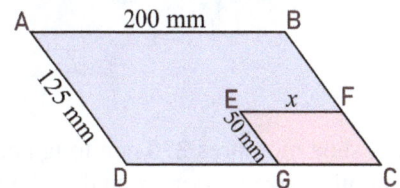

$\dfrac{125}{200} = \dfrac{50}{x}$	This is the proportion.
$125x = 50 \cdot 200$	After cross-multiplying, we get this equation.
$125x = 10\,000$	The next step is to divide both sides of the equation by 125.
$x = 80$	This is the final answer. So, EF = 80 mm.

You may use a calculator for all the problems in this lesson.

1. Triangle ABC was dilated to produce triangle DEC.

   a. What is the centre of the dilation?

   b. What is the scale factor of the dilation?

   c. What is the scale *ratio* between △ABC and △DEC?

   d. If BC = 58 cm, how long is CE?

   e. If CD = 75 cm, how long is AC?

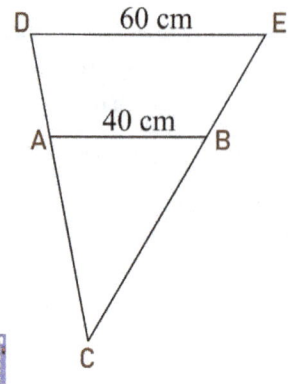

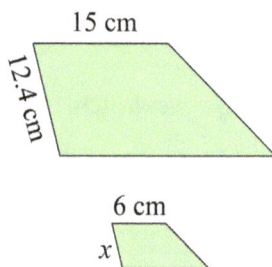

2. The larger trapezoid on the left is dilated as shown (it shrinks).

   a. What can you say about the corresponding angles in both figures?

   b. What is the scale factor?

   c. What is the scale *ratio* (in lowest terms)?

   d. Calculate *x*, to the nearest tenth of a centimetre.

3. The two triangles are similar. Notice carefully which sides are corresponding sides. Then calculate the lengths of the sides marked with *x* and *y*.

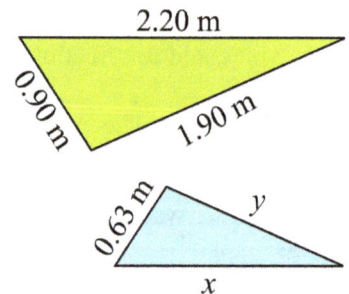

4. A shoe measures 270 mm long and 94 mm wide. The same kind of shoe in another size is proportional to the other shoe. If the smaller shoe is 245 mm long, how wide is it?

Similar Figures: More Practice

You may use a calculator for all the problems in this lesson.

1. **a.** Prove that the two figures are similar by giving a sequence
   of transformations (with details) that maps one to the other.

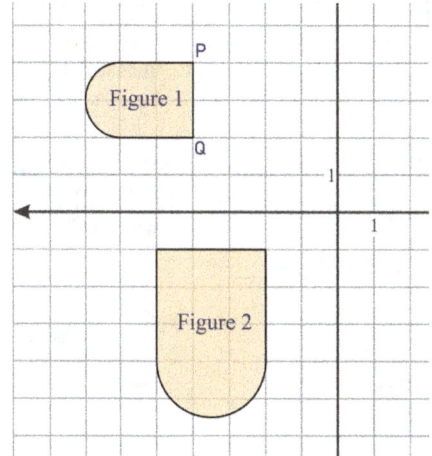

   b. What is the scale ratio between figure 1 and figure 2?

   c. What is the scale factor?

   d. Figure 1 consists of a semicircle and a square. Calculate its perimeter.

$C = \pi \cdot d$
(C is the circumference and *d* the diameter of the circle.)

   e. Now use the scale ratio to find the perimeter of figure 2.

2. Triangle PQR was dilated from point R to map to
   triangle P'Q'R.

   a. Calculate P'Q'.

   b. Given that PR is 5.83 units, calculate P'R.

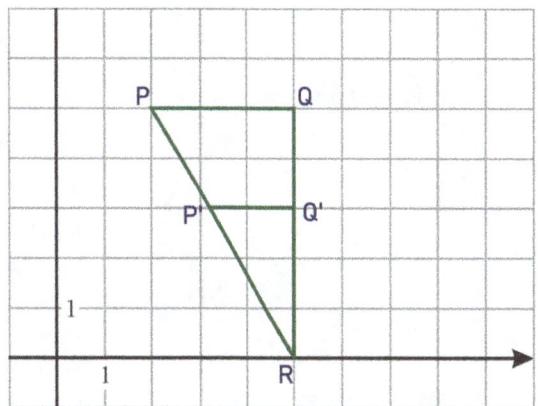

Note on notation. When two angles are **marked with a single arc**, with no measurement or letter next to the arc, it signifies they are **congruent**. The same is true of a double or triple arc.

In this illustration, we have two similar triangles. Their corresponding angles are congruent (have the same angle measure).

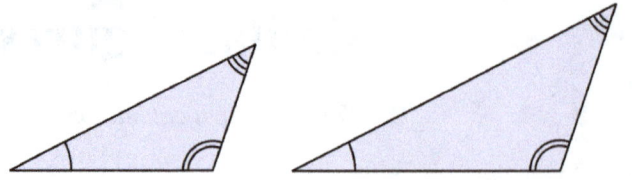

3. All the triangles depicted are similar. Find the unknown side lengths.

$w =$ _____

$x =$ _____

$y =$ _____

$z =$ _____

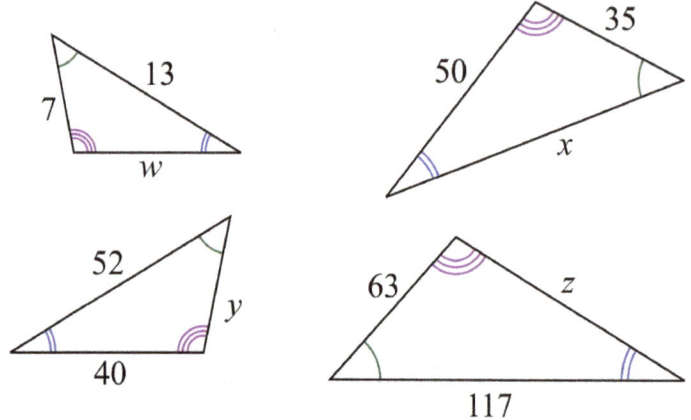

4. The two quadrilaterals are similar.

   a. Calculate the side lengths marked with x and y, to the nearest tenth of a centimetre.

   b. (challenge) Find the value of z.

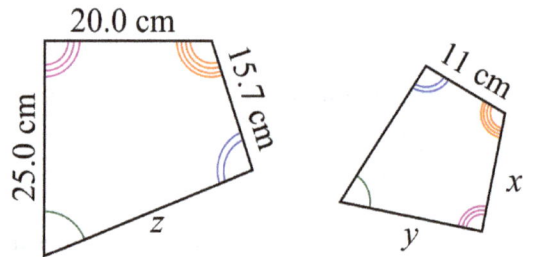

5. Which statement is true?

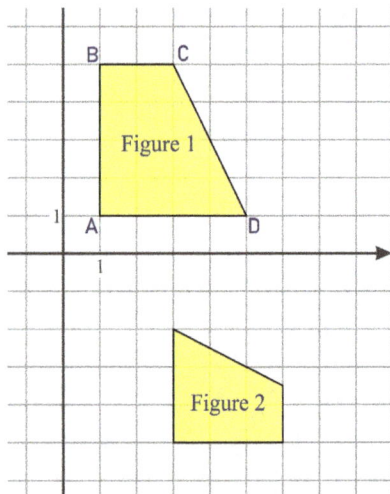

   a. A reflection in the x-axis, followed by a 90-degree rotation counterclockwise around point B', followed by a translation 6 units to the right, followed by a dilation from point B''' with scale factor 3/4 will transform Figure 1 to Figure 2, so they are similar figures.

   b. There is no sequence of transformations that would map figure ABCD to the other, so the figures are neither congruent nor similar.

   c. A dilation from point A with scale factor 3/4, followed by a 90-degree rotation counterclockwise around point A', followed by a reflection in the vertical line at $x = 1$, followed by a translation 6 units down and 2 to the right will transform Figure 1 to Figure 2, so they are similar figures.

6. Which <u>two</u> of the following figures are similar?

   (i) A parallelogram with angles 50°, 130°, 50°, and 130°, and side lengths 9 cm and 4 cm.

   (ii) A parallelogram with angles 50°, 130°, 50°, and 130°, and side lengths 27 cm and 16 cm.

   (iii). A parallelogram with angles 30°, 150°, 30°, and 150°, and side lengths 18 cm and 8 cm.

   (iv) A parallelogram with angles 50°, 130°, 50°, and 130°, and side lengths 72 cm and 32 cm.

7. A rectangle with sides 3 units and 5 units is dilated by scale factor 2.
   How is its area affected?

8. Consider the triangle on the right with the altitude h = 6 units and base b = 11 units.

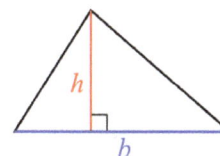

   a. If this triangle is dilated with scale factor of 2, how long is the altitude of
   the dilated triangle? Its base?

   b. How does the area of the dilated triangle relate to the area of the original triangle?

Puzzle Corner Investigate how the area of a rectangle changes when it is dilated. Make up some example rectangles (decide their width and height), choose some simple numbers for scale factors, and make a table so you can organize your findings. For example, you might check what happens with scale factors 2, 3, 4, 10, and 1/2.

Original rectangle			Scale Factor	Dilated rectangle		
Width	**Height**	**Area**		**Width**	**Height**	**Area**

Review: Angle Relationships

A **ray** has a starting point and continues indefinitely in one direction (indicated by one arrowhead).

An **angle** consists of **two rays that start at the same point**, called the **vertex**. Each ray is called a **side** of the angle.

We can denote the angle on the right as ∠BAC (angle BAC) or as ∠CAB. Note that the vertex point is listed in the middle. We could also label the angle by giving it a name (α, or alpha, in this case).

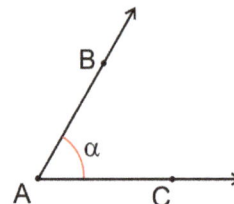

Two angles are **adjacent** if they have a **common vertex and share one side (a ray).**

In the image on the right, ∠α and ∠β (beta) are adjacent angles.

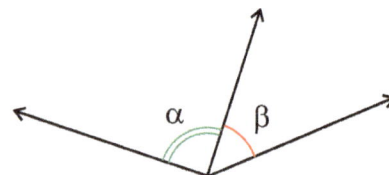

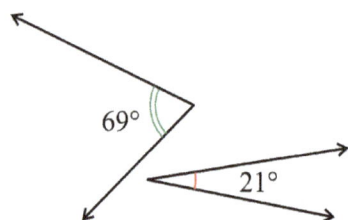

← Two angles are **complementary** if their angle sum is 90 degrees.

Two angles are **supplementary** if their angle sum is 180 degrees. →

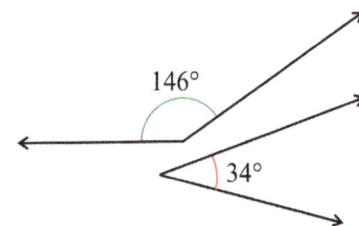

Supplementary angles don't have to be adjacent, and neither do complementary angles — but they often are, in various geometric figures.

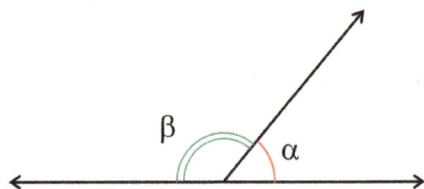

The angles ∠α and ∠β in this image are adjacent supplementary angles (also called a linear pair).

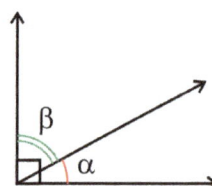

The angles ∠α and ∠β in this image are adjacent complementary angles.

Here's a mnemonic to help you remember complementary and supplementary angles:
Supplementary angles form a **S**traight line, and **C**omplementary angles form a **C**orner (a right angle). (Or, Since C < S in the alphabet, C is the 90 and S is the 180).

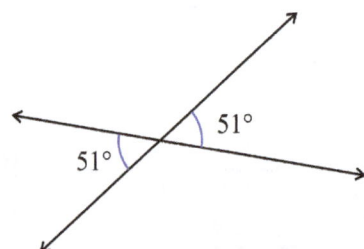

When two lines intersect, they form four angles. The two opposite angles are called **vertical angles**.

Vertical angles are **congruent**. (They have the same angle measure.)

1. The figure shows a line and a ray, forming two supplementary angles. Find the value of x.

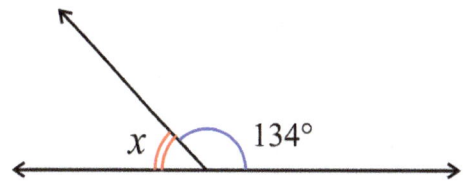

x 134°

2. Two lines intersect as shown in the diagram, forming four angles.

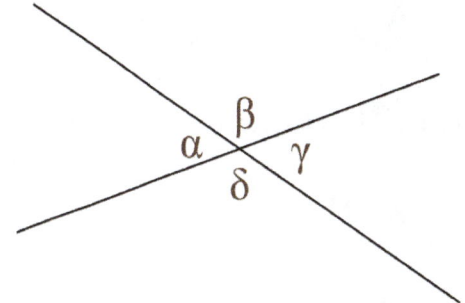

   a. How many angles, at a minimum, would you need to measure in the diagram, in order to be able to calculate the rest?

   b. Find the angle measures of all four angles in the diagram
   (alpha, beta, gamma, delta — the first four letters of the Greek alphabet).

   $\angle\alpha =$ _____° $\angle\beta =$ _____°

   $\angle\gamma =$ _____° $\angle\delta =$ _____°

3. Find the unknown angle measures x and y. (But do not measure.)

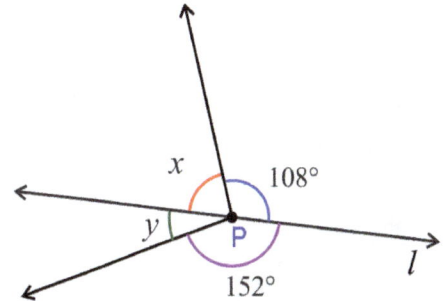

x 108°
y P l
152°

4. **a.** Measure the angles in the parallelograms below. Based on your findings, fill in the words "complementary", "supplementary", or "congruent" on the empty lines.

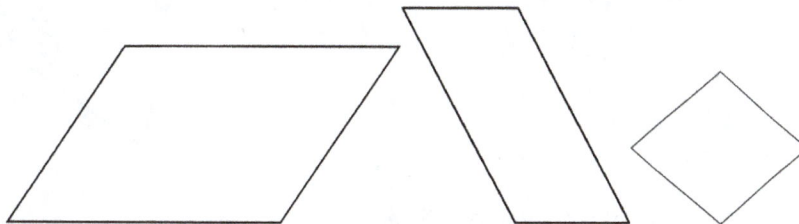

   Adjacent (neighbouring) angles in a parallelogram are _____

   Opposite angles in a parallelogram are _____

   b. What is the angle sum in a parallelogram?

5. Find the angle measures of ∠α, ∠β, and ∠γ.

∠α = _____°   ∠β = _____°

∠γ = _____°

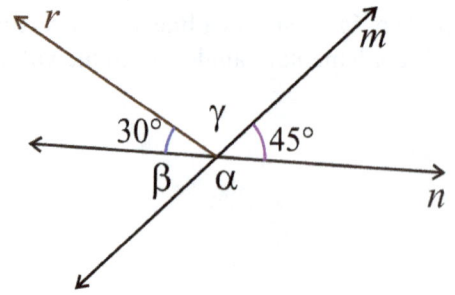

6. Two lines intersect. Find the value of x.

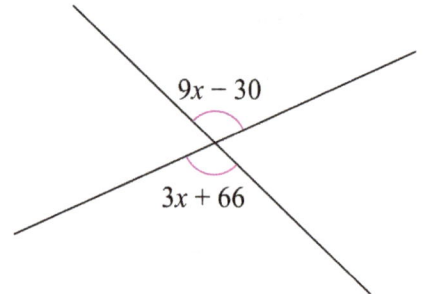

7. A right triangle has one right angle. What special relationship exists between the other two angles? (Use vocabulary from this lesson.) You can measure to find out.

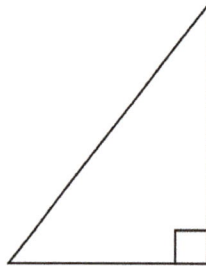
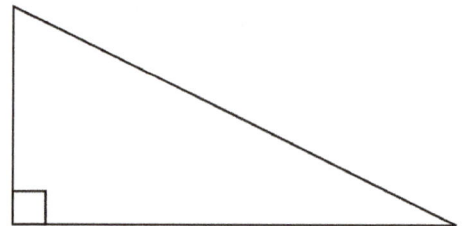

8. Several rays start from the same point. Solve for the unknown x.

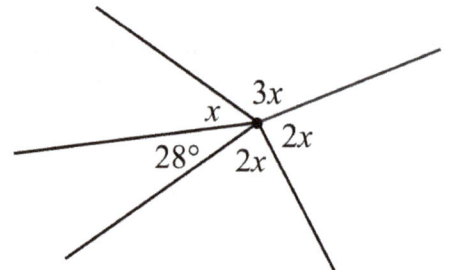

9. The midpoints of each of the sides of rectangle ABCD are connected with line segments to form a quadrilateral.

   a. What type of quadrilateral is it?

   b. There are four triangles formed. What kind of triangles are they (obtuse, acute, right, scalene, equilateral, isosceles)?

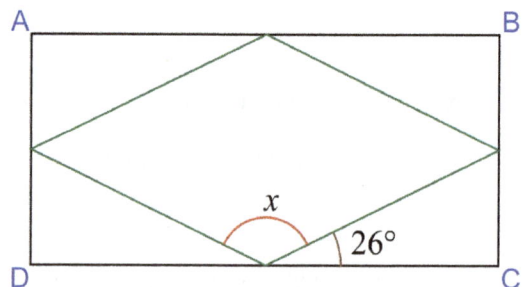

   c. Find angle x. (Do not measure.)

Corresponding Angles

Here you see two lines, and a third line, called a **transversal,** that intersects them both.

In this lesson we will examine the various angles formed by this arrangement, in particular when the two lines are parallel.

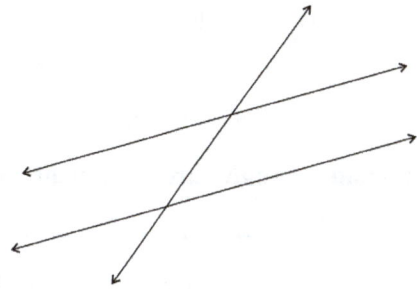

1. In the illustration below, two parallel lines are cut by a transversal.

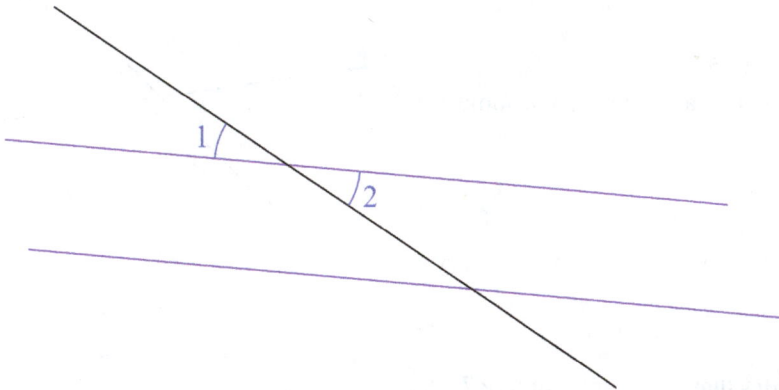

    a. Find four pairs of vertical angles in the figure. One is already marked with numbers 1 & 2. Mark the others with 3 & 4, 5 & 6, and 7 & 8.

    b. Find the angle measures of the eight angles. (You can measure; however, you don't need to measure all of them!) Which angles are congruent?

2. In this illustration, two lines are again intersected by a transversal, but this time the two lines are not parallel.

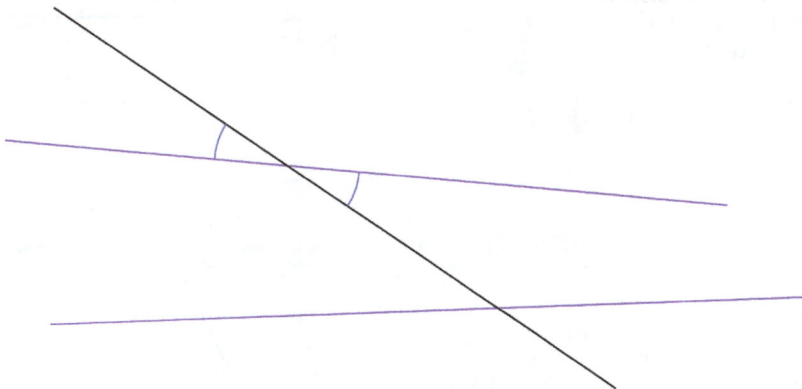

    How does this situation compare with the one in question #1? What is the same? What is different?

Lines L_1 and L_2 are parallel and transversal T intersects them both.

Angles 1 and 2 are **corresponding angles**. This means they are on the same side of the transversal, and on the same side of L_1 and L_2.

Angles 3 and 4 are also corresponding angles.

Corresponding angles are congruent.

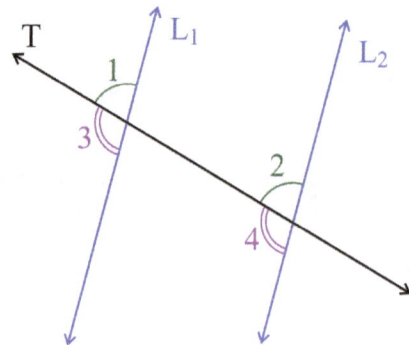

3. Two parallel lines are cut by a transversal.

   a. Which angle is a corresponding angle to $\angle 2$?

   b. Which angle is a corresponding angle to $\angle 8$?

   c. If $\angle 7$ is 127°, what is $\angle 5$?
   How do you know? (Use appropriate vocabulary.)

   d. What is the measure of $\angle 6$?
   How do you know? (Use appropriate vocabulary.)

   e. Identify all the angles that are congruent to $\angle 7$.

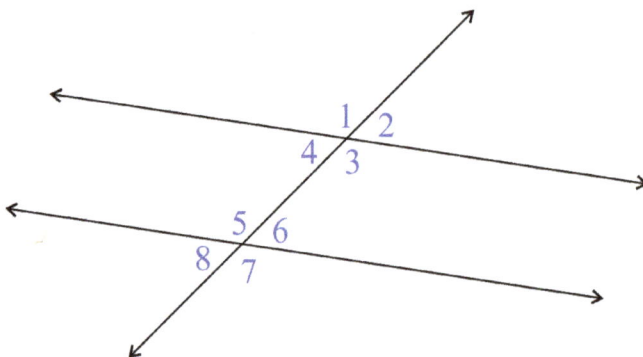

4. Lines m and n are parallel. Show that $\angle ABC$ is congruent to $\angle \alpha$ using a rigid transformation.
   (In other words, give detailed information about the transformation that would map $\angle ABC$ to $\angle \alpha$.)

   (Remember, *rigid transformations* are those that preserve distances, angles, and parallel lines: translations, rotations, and reflections.)

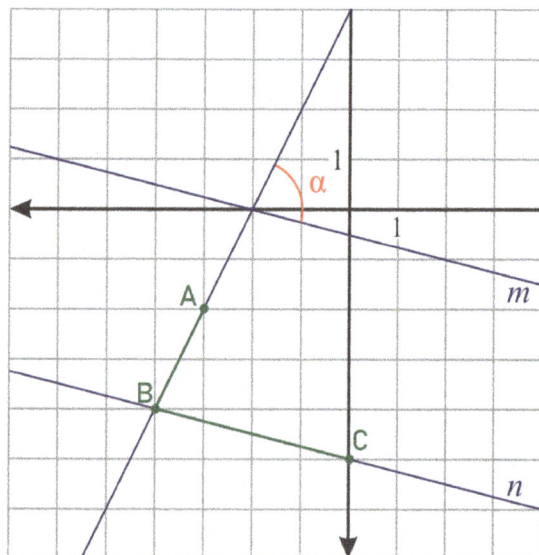

More Angle Relationships with Parallel Lines

1. Two parallel lines are cut by a transversal.
Point M is the midpoint of $\overline{AB}$.

How could you use a rigid transformation to
prove that angle 2 is congruent to angle 1?

You can use transparent paper to help you
investigate this.

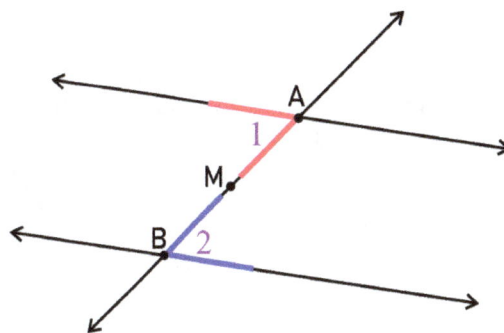

Angles 1 and 2 are called **alternate interior angles**. They are on alternate sides of the transversal
and in between the two parallel lines — in an "inner" position in relation to the whole diagram.

Angles 3 and 4 are also alternate interior angles. **Alternate interior angles are congruent.**

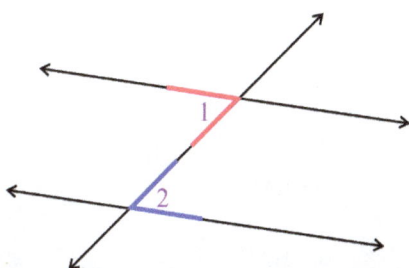

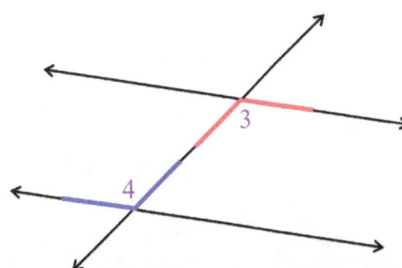

Angles 5 and 8 are **alternate exterior angles**.
They are on alternate sides of the transversal and
in an "outer" position in relation to the whole diagram.

Angles 6 and 7 are also alternate exterior angles.

Alternate exterior angles are congruent.

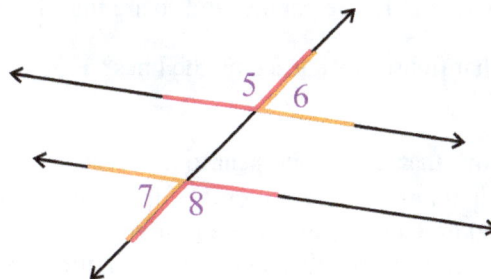

2. Lines L_1 and L_2 are parallel. Fill in the blanks, describing the
types of angles formed.

Angles 5 and 7 are _____ angles.

Angles 3 and 5 are _____ angles.

Angles 1 and 7 are _____ angles.

Angles 2 and 6 are _____ angles.

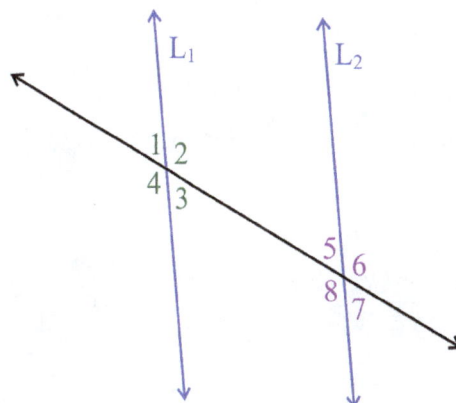

3. Two parallel lines are cut by a transversal. Point M is the midpoint of $\overline{CD}$.

   a. What is the relationship between ∠1 and ∠2?

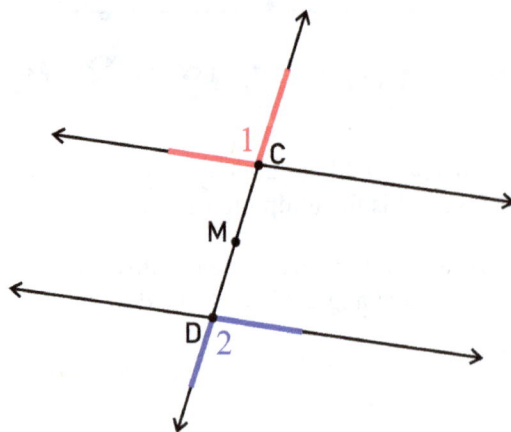

   b. How could you use a rigid transformation to prove that ∠1 is congruent to ∠2?

4. Lines L_1 and L_2 are parallel and ∠3 = 53°.

   a. What is the measure of angle 5?

   How do you know?

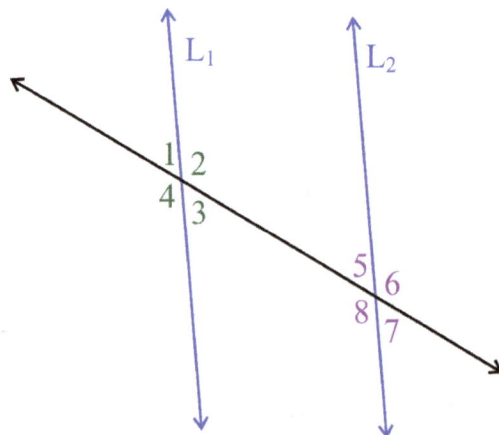

   b. What is the measure of angle 8?

   How do you know?

5. Lines L_1 and L_2 are parallel and so are lines L_3 and L_4.

   a. What figure is enclosed by the lines?

   b. Prove that ∠1 is congruent to ∠3. You might need to refer to some of the angles 5, 6, 7, and/or 8 so they are marked for you. You may certainly mark other angles in the figure, too, so you can refer to them in your proof.

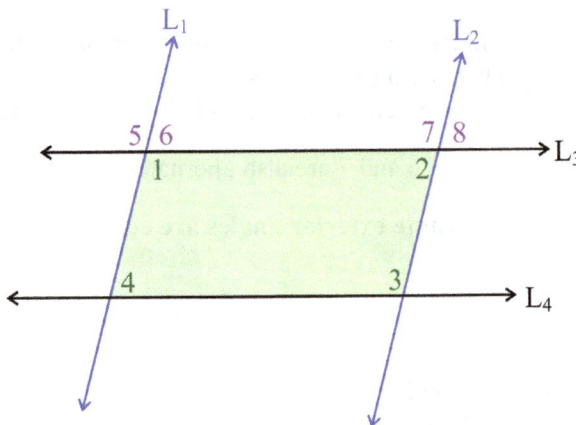

The Angle Sum of a Triangle

1. Draw as many triangles as needed so you have at least one of each: obtuse, acute, right, scalene, isosceles and equilateral. Think whether you really need to draw six different ones, or maybe you can get by with less. Once drawn, measure their angles and calculate the angle sum of each. What do you observe? Make a conjecture.

2. We will now study a neat proof about the angle sum of a triangle.

   Here we see triangle ABC and line l that is parallel to $\overline{AC}$.

   The proof is based on finding three adjacent angles that make a straight angle (180°) along a line, and that are congruent to the angles in the triangle.

   Perhaps you can see how the proof works by studying the diagram. Below are some hints.

   a. Find angles in the diagram that are congruent to ∠A.
      Mark them with a single arc.

   b. Find angles in the diagram that are congruent to ∠C.
      Mark them with a double arc.

   c. Find an angle in the diagram that is congruent to ∠B.
      Mark it with a triple arc.

   d. Can you now find three adjacent angles that form a straight angle together, and that are congruent to the three angles in △ABC?

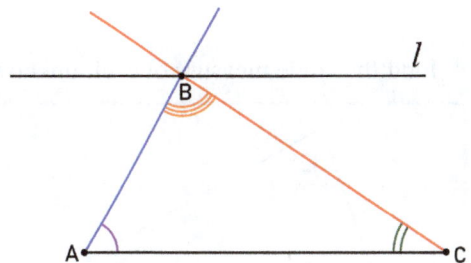

The sum of the angles in a triangle is 180°.

$$a + b + c = 180°$$

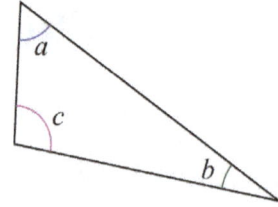

3. Fill in the missing parts in the proof for the fact that the angle sum of a triangle is 180°.

In the diagram, we see triangle ABC, and its angles α, β, and γ.
The line *l* is drawn so that it is parallel to $\overline{AC}$.

Angles α and α′ are congruent because they are

_____ angles.

Angles β and β′ are _____ because they

are vertical angles.

And angles γ and γ′ are congruent because they are, again,

_____ angles.

Therefore, the sum ∠α′ + ∠β′ + ∠γ′ is equal to the sum ∠α + ∠β + ∠γ.

Since the three angles α′, β′, and γ′ are adjacent and form a _____ angle, the sum of their angle

measures is _____°. This means that ∠α + ∠β + ∠γ = _____°, too.

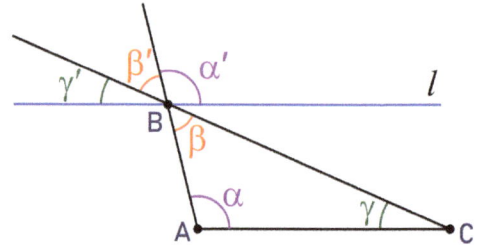

4. Find the angle measure of each unknown angle. *Do not measure.*

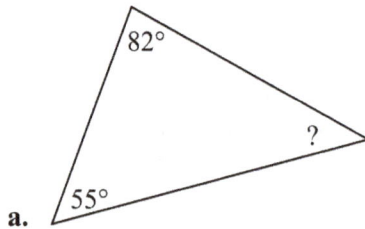

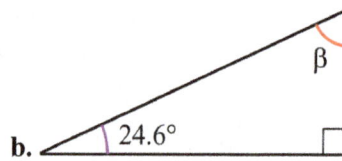

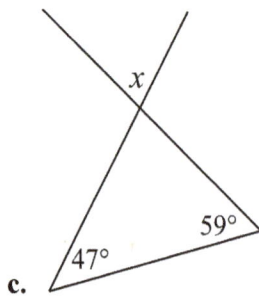

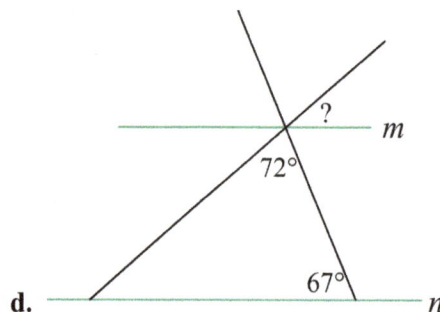

a. (triangle with 82°, 55°, and ?)

b. (right triangle with 24.6° and β)

c. (crossing lines with x, 47°, 59°)

d. (crossing lines with parallel lines m and n, 72°, 67°, ?)

98

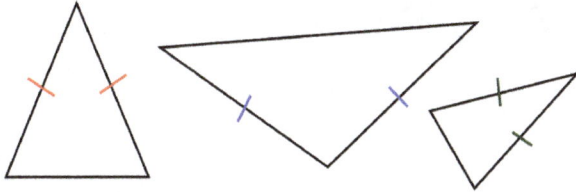

An **isosceles triangle** has at least two sides that are congruent (the same length; marked with a single mark to signify congruency). Think of it as a "same-legged" triangle, the "legs" being the two sides that are the same length.

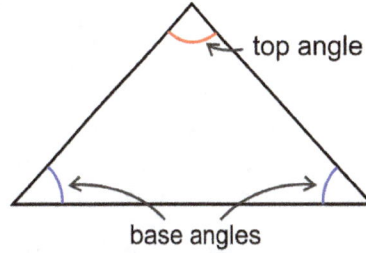

Not only does an isosceles triangle have two congruent sides, but two of its angles are also congruent. They are called the **base angles**. The remaining angle is called the **top angle**.

5. **a**. The top angle of an isosceles triangle measures 24°.
   What do the base angles measure?

   b. The base angles of an isosceles triangle measure 73°.
   What does the top angle measure?

6. Is an equilateral triangle isosceles?

7. An equilateral triangle is also an equi*angular* triangle, meaning that all its angles are congruent.
   What is the measure of each of its angles?

8. Can you draw a triangle that has two obtuse angles?
   Why or why not?

9. Draw an isosceles triangle with a 50° top angle
   and two 6.4 cm sides.
   How long is its third side?
   What do its base angles measure?

Exterior Angles of a Triangle

If you extend any of the sides of a triangle, a new angle is formed adjacent to one of the interior angles of the triangle. It is called **an exterior angle** of the triangle.

The illustration shows two such exterior angles. We would find four more if we also extended the other two sides of the triangle. Each triangle actually has six exterior angles.

If you know the measure of the angle in the triangle, you can calculate the exterior angle, since the two are supplementary angles.

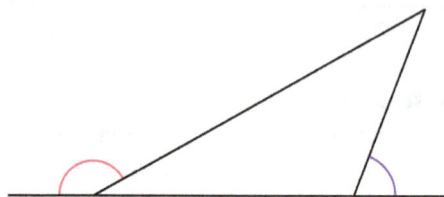

1. In each picture, one or more sides of the triangle are extended in a straight line. Find the angle measure of each unknown angle. *Do not measure.*

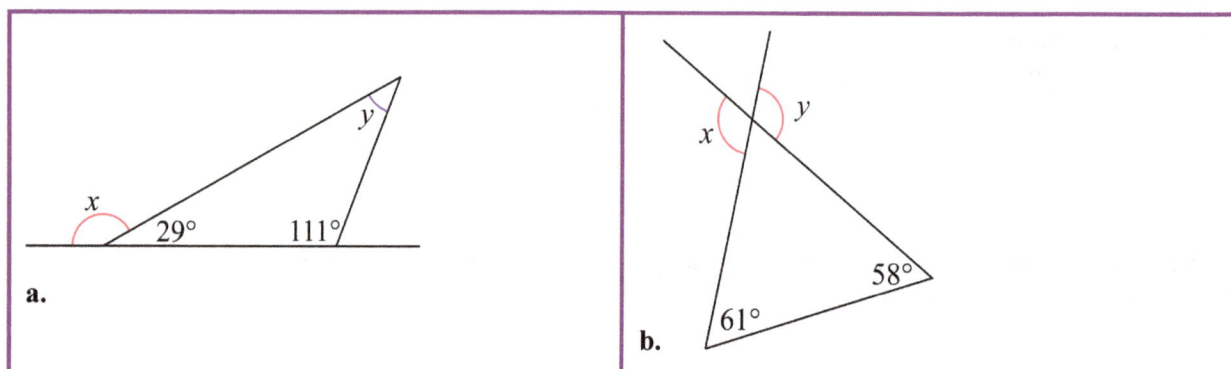

a.

b.

2. Find the measure of angle z. Explain and justify your reasoning. Angles G, H, and J are named in the picture. You can name more angles if necessary.

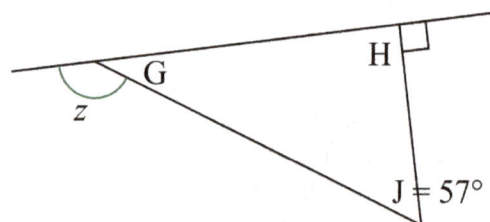

3. Extend the sides of the triangle, and mark all six of its exterior angles.

   Remember, each exterior angle is supplementary to one of the interior angles of the triangle.

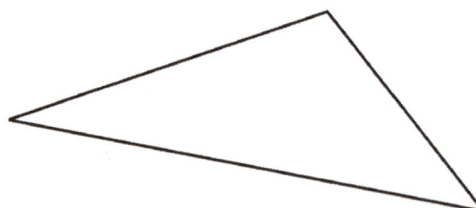

100

4. Find the measure of angle x. Justify your reasoning.

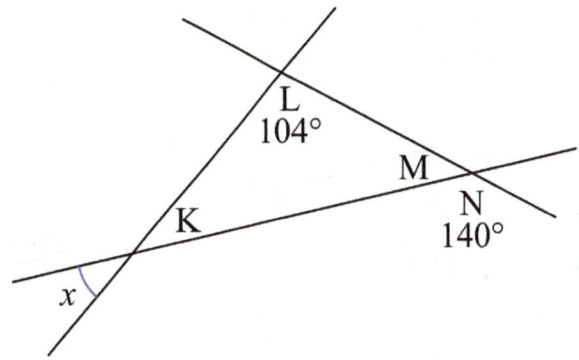

5. Find the value of each unknown, using an equation.

a.

$3x + 7$

$x + 23$

x

b.

$2y - 51$

$y + 32$

$y - 29$

c.

$2w + 53$

$w - 15$

$4w$

6. Find the value of *x*, in each case. Show your work.

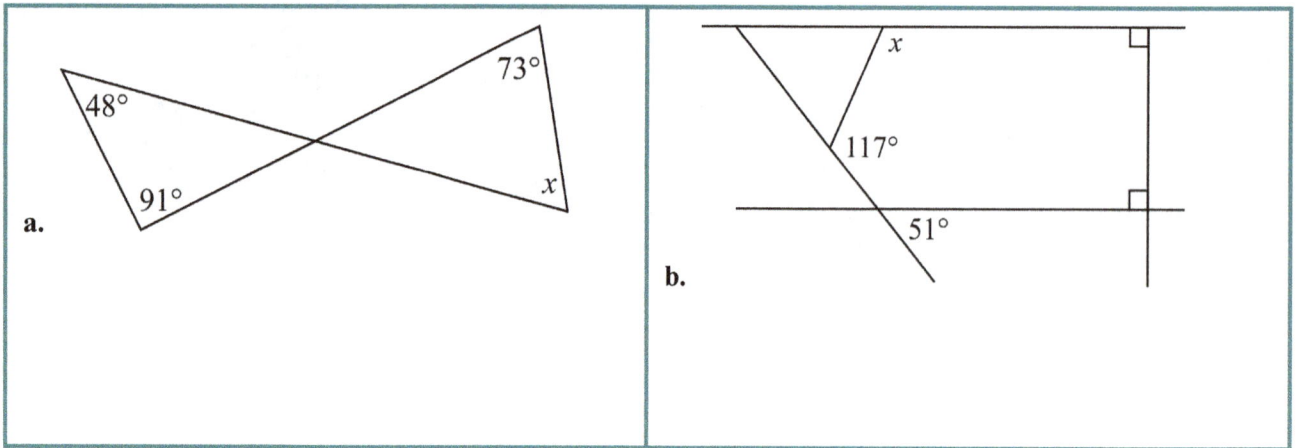

a.

b.

7. In your own words, prove that the angle sum of a triangle is 180°.
   Use the given diagram (you can add to it).

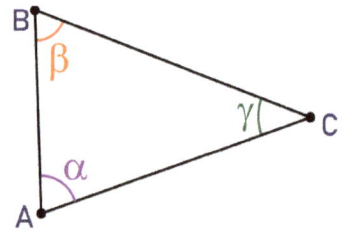

8. Find the value of *x*, and show your work.

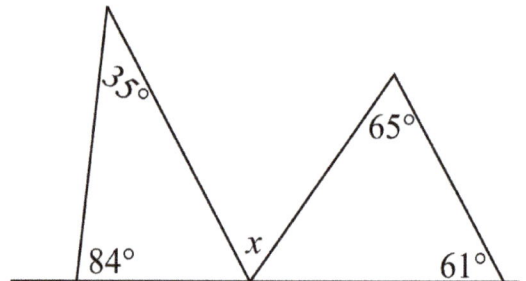

Any quadrilateral can be divided into two triangles by drawing a diagonal. We can use this idea to easily prove what the sum of the angles of any quadrilateral must be.

a. What is the sum of the angles of any quadrilateral?

b. Prove it.

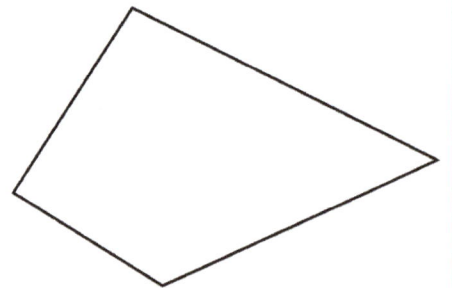

Puzzle Corner

Angles in Similar Triangles, Part 1

> Recall that a dilation preserves angles, and the distances between points are multiplied by the scale factor of the dilation.
>
> This means that when two figures are similar...
>
> - their corresponding angles are congruent (equal).
> - their corresponding sides are proportional; in other words in the same ratio.

> When it comes to *triangles*, something stronger is true: we don't even need to check the sides. If the three angles of a triangle are congruent to the three angles of another triangle, that alone establishes that <u>the two are similar triangles</u>.

1. Are these two triangles similar?

   How do you know?

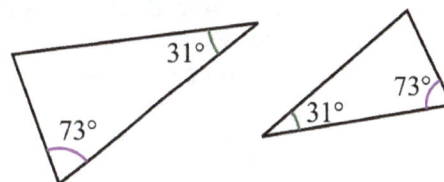

2. Two of the angles of triangle T are equal to two of the angles of triangle S.

   a. Can you conclude, based on this information, that triangles T and S are similar? Why or why not?

   b. Can you conclude, based on this information, that triangles T and S are congruent? Why or why not?

3. **a.** Using a protractor, draw (on a separate paper) two triangles, of different sizes, so that two of the angles of the one triangle are congruent to two of the angles of the other. If you have drawn accurately, having two congruent angles means your triangles should be similar. (Can you see why?)

   b. Measure the sides of your triangles, write the ratios of the corresponding sides below, and calculate their values to three decimal digits. (There will be three such ratios.) Verify that the corresponding sides are in the same ratio

   <u>Ratios:</u>

   Keep in mind your measurements can only be within the nearest millimetre, not fully exact, and thus, both your measurements and the ratios you calculate will only be approximations.

   c. (optional) Calculate the two scale factors involved: one you would use if going from your smaller triangle to your larger, and the other, going from your larger triangle to the smaller. Use the *average* of the three ratios of the corresponding sides.

   d. (optional) Verify that the two scale factors are reciprocals (i.e. their product is one, or in this case, very close to one, since we are using measurements that are, by their very nature, rounded and not accurate).

The AA criterion for similar triangles

If two angles of a triangle are congruent to two angles in another triangle, the triangles are similar.

This is called the "AA" criterion — the two A's stand for **Angle** + **Angle** (or two angles) being congruent.

4. Are the triangles similar? Explain how you know.

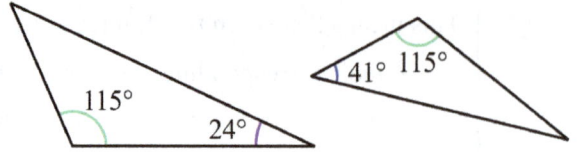

5. $\overline{AC}$ is parallel to $\overline{A'C'}$.

   a. Is △ABC similar to △A'BC'? Explain how you know.
   (If you do not see it, extend line AC and use what
   we learned about parallel lines and transversals.)

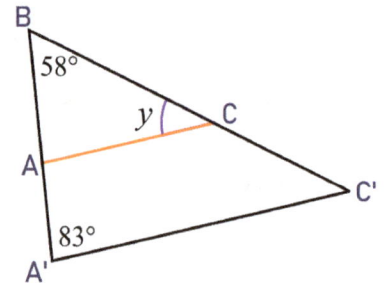

   b. What is the value of y?
   Show your work.

6. Figure ABCD is a parallelogram.
   What is the value of x?

   Explain your reasoning.

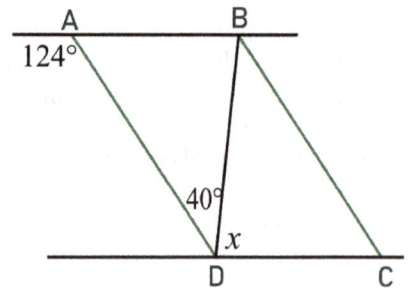

7. $\overline{AB}$ is parallel to line l. What is the value of x? Explain.

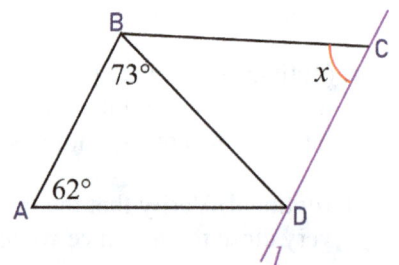

104

Angles in Similar Triangles, Part 2

1. Two lines intersect at point K.

   a. Are these similar triangles? Explain!

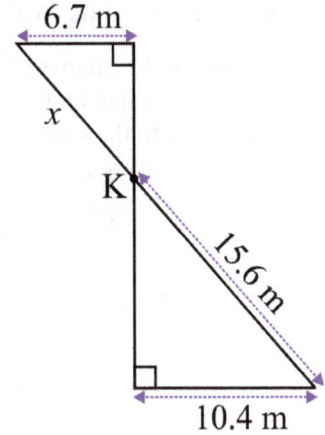

   b. Find the value of x.

2. Are the two triangles similar?

   Explain how you know.

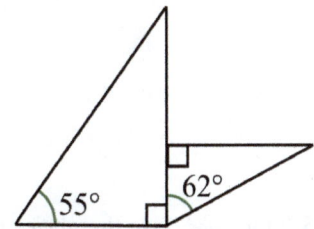

3. (Challenge) Lines m and n are parallel.
   Are the two triangles similar?
   Explain how you know.

   (Feel free to mark angles in the image
   so you can reference them.)

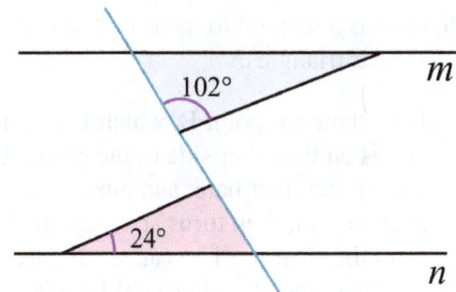

4. Here is a method to measure the height of tall objects, using shadows.

   (i) Measure your height and the length of your shadow.

   (ii) At the same time (without waiting much), measure the shadow of the tall object, such as a tree.

   (iii) Look at the images. As long as the sun is in the same position in the sky, triangles ABC and ADE will have the same angles, which means they are similar. This, in turn, means that the ratio of the shadows is equal to the ratio of the heights of the two objects. This allows you to calculate the height of the tree (e.g. by using a proportion).

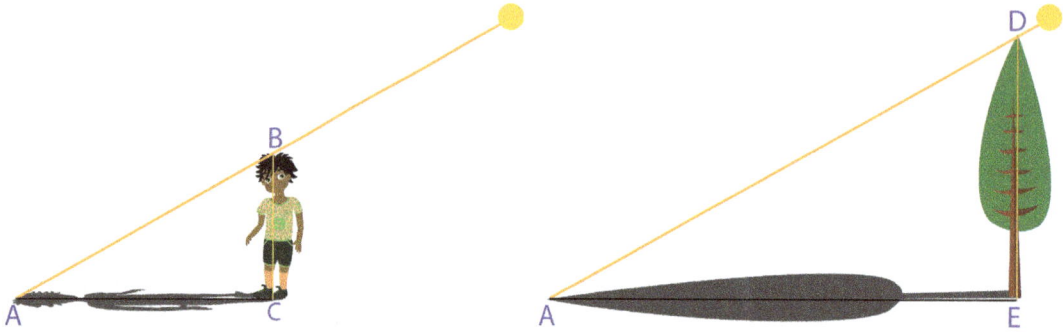

The boy is 145 cm tall, his shadow is 207 cm long, and the shadow of the tree is 600 cm. Calculate the height of the tree to the nearest centimetre.

5. *Optional.* Use the above method to find the height of some tall objects and/or trees in the neighbourhood.

6. Here is a method to measure the width of a river using similar triangles.

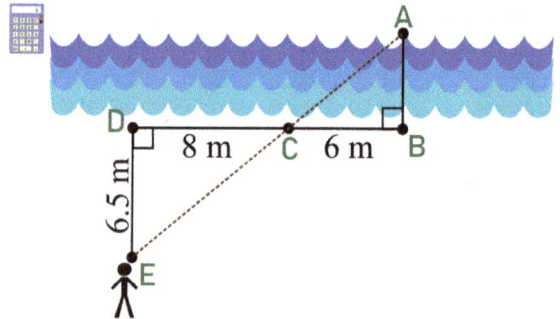

John stands at point B, which is directly across from a tree that is on the other side of the river, at A. Then he walks 6 m on the river bank and puts a stake (at C). He walks another 8 m, then turns at a right angle and walks away from the river until he can see the tree and the stake at the same line of sight (point E). He then measures DE — it is 6.5 m.

Find the width of the river to the nearest tenth of a metre.

Volume of Prisms and Cylinders

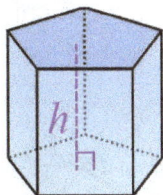

A pentagonal prism. The dashed line marks the height (h).

In geometry, a **prism** is a 3-dimensional solid with two identical parallel polygon faces, known as its **bases**, and the rest of the faces are rectangles (or parallelograms, if the prism is oblique, or "slanted"). In this book we will only deal with right prisms (ones that are not slanted).

Note that a box with a rectangular base is a prism; in geometry we call it a *rectangular* prism.

A **cylinder** is similar to a prism, but its bases are circles or ellipses, and it has just one other face that is wrapped around the bases.

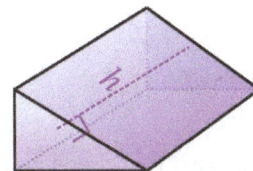

A triangular prism. Note that its bottom face is facing the viewer. The dashed line marks the height (h).

Both prisms and cylinders are named after the polygon or shape at their base.

The **height** of both prisms and cylinders is the length of the line segment drawn from the top face to the bottom face that is perpendicular to both faces.

The **volume** of prisms and cylinders is calculated in the same way: we simply **multiply the area of the base (A_b) by the height** (h).

The formula is: $V = A_b h$.

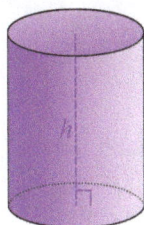

A circular cylinder

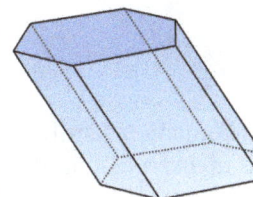

An oblique hexagonal prism

Example 1. Calculate the volume of this cylinder.

This cylinder is "lying down." Imagine standing it up to see what the top and bottom faces are. Its top and bottom faces are circles.

Notice that we are given the *diameter* of the circle, but to calculate the area of a circle, we need to use the *radius*, which is 1.1 inches.

Using the formula for the area of a circle, $A = \pi r^2$, we get that the area of the bottom face is $\pi \cdot (1.1 \text{ in})^2$. The height is 7.5 in.
The volume is the product of the two: $V = \pi \cdot (1.1 \text{ in})^2 \cdot 7.5 \text{ in} \approx \underline{28.5 \text{ in}^3}$.

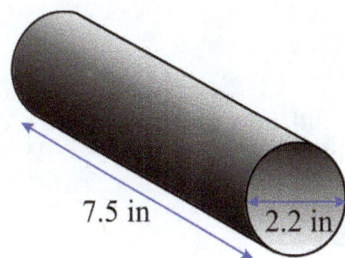

7.5 in   2.2 in

You can use a calculator in all the problems in this lesson.

1. You have learned to calculate the volume of a box (a rectangular prism) by multiplying its width, depth, and height (its three dimensions).

   Does the formula $V = A_b h$ also apply to boxes?

   Why or why not?

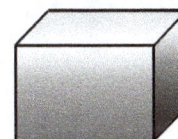

2. The bottom face of this cylinder is a circle with a diameter of 8 cm. Its height is 12 cm.
   Find its volume to the nearest ten cubic centimetres.

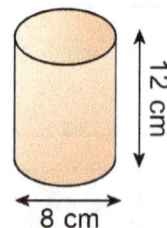

3. **a.** What is this shape called? If you are unsure, ask yourself:
   what are the two identical parallel faces?

   b. Calculate its volume.

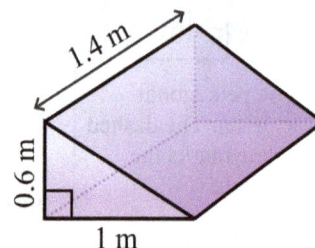

4. The Fernandez family has three cylindrical water tanks, of different sizes. The first one has a diameter of
   1.52 m and a height of 1.8 m. The second and third have a diameter of 2.4 m and a height of 3.0 m.

   a. Calculate their total volume in cubic metres.

   b. The family uses 450 litres of water per day, on average. If the water tanks are full, how many days of
   water supply do they provide for the family? Note: One cubic metre = 1000 litres.

5. Find a drinking cup or a mug with a cylindrical shape. Most drinking glasses taper down towards the bottom
   so they don't work. Look for one whose bottom and top faces are congruent circles.

   a. Measure the mug, and calculate its volume in cubic centimetres.

   b. Measure its volume now in millilitres, using a measuring cup, and compare to what you got above.
      Remember that 1 ml = 1 cm^3.

   If the results are far apart, check your measurements. Check also whether your measuring cup is accurate
   (often they are not).

Volume of Pyramids and Cones

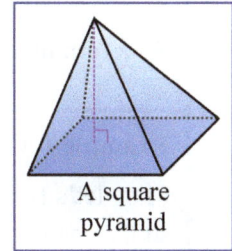

A pyramid is a solid that has some polygon as a base. Its other faces are triangles that meet at the top vertex of the pyramid.

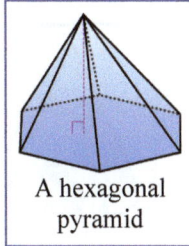

A hexagonal pyramid

Just like prisms, pyramids also are named after the polygon at their base. A rectangular pyramid has a rectangle as its base, a triangular pyramid has a triangle as its base, a pentagonal pyramid has a pentagon as its base, and so on.

A square pyramid

The *height* or *altitude* of a pyramid is the length of the line segment drawn from the top vertex to the base so that it is perpendicular to the base. We need the height in calculating the volume of pyramids.

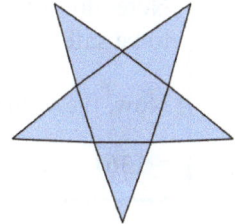

Can you tell what kind of pyramid the net on the right belongs to?

You can find the answer below this blue box, but think first!

A cone is similar to a pyramid, but it has a rounded shape as its base. The cone on the right is a circular cone. And similarly with pyramids, a cone has a *height*: a line drawn from the vertex that is perpendicular to the base.

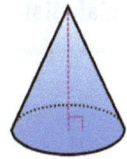

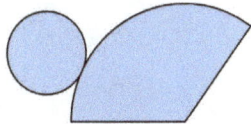

The net of a cone has two parts: a circle (the base), and a sector (a part) of a circle, which is the other face of the cone — the one you wrap around the base.

Note: The net above is for a pentagonal pyramid: it has a pentagon as a base, and triangles as the other faces.

You may use a calculator in all problems in this lesson.

1. Name the solids.

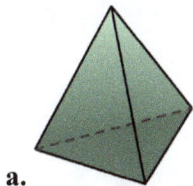

a.

b.

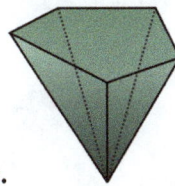

c.

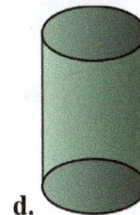

d.

2. Name the solids that can be constructed from these nets.

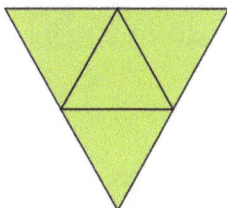

a.

b.

c.

d.

The **volume** of all pyramids and cones is calculated in the same way: It is one-third of the area of the base (A_b) multiplied by the height (h). It doesn't matter whether the cone or the pyramid is slanted or upright; the formula works in either case.

As a formula, we write $V = \dfrac{1}{3} A_b h$.

Example 1. Calculate the volume of the cone to the nearest ten cubic centimetres.

First, let's find out the area of the base. It is a circle with a radius of 8.5 cm, so its area is $A_b = \pi \cdot (8.5 \text{ cm})^2 \approx 226.865 \text{ cm}^2$.

Note: don't round your intermediate answers a lot. Keep a few extra digits just to be safe. Rounding to the nearest ten should only happen in the final step.

18 cm

17 cm

Now, the volume. Using the formula, we get $V = \dfrac{1}{3} A_b h = \dfrac{1}{3} \cdot 226.865 \text{ cm}^2 \cdot 18 \text{ cm}$

$= 1361.19 \text{ cm}^3$ or approximately <u>1360 cm^3</u>.

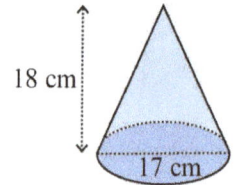

3. Calculate the volumes of these solids. Note: the cones are circular (have a circle as their base).

a.
18 cm
21 cm^2

b.
2.0 m
2.0 m
1.5 m

c.
2 cm
7 cm

d.
8.5 cm
9.5 cm

e.
9 cm
10 cm
10 cm
an octahedron:
two square pyramids

f.
6.0 cm
3.8 cm
6.6 cm
a circular cone and
a circular cylinder

Notice something similar about the two formulas for volume that we have studied:

Volume of a prism or cylinder: $V = A_b h$.

Volume of a pyramid or cone: $V = (1/3)A_b h$.

This means that if we take a pyramid and a prism with the <u>same base and same height</u>, the volume of the pyramid is exactly one-third of the volume of the prism. The same is true of a cone and a cylinder with the same base and same height.

This relationship might remind you of something similar concerning areas: the area of a triangle is always 1/2 of the area of a parallelogram with the same base and height!

A pyramid inside a box:
(The pyramid and the prism share the same base and the same height.)

The volume of the pyramid is 1/3 of the volume of the box.

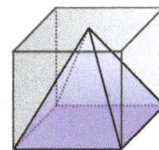

A cone inside a cylinder:
(The cone and the cylinder share the same base and the same height.)

The volume of the cone is 1/3 of the volume of the cylinder.

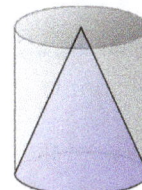

4. A cube with a volume of 27 000 cm^3 has a square pyramid inside it so that the base of the pyramid is the same as the base of the cube, and its vertex touches the top of the cube.

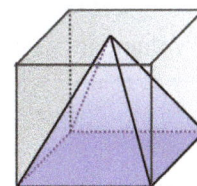

   a. What is the volume of the pyramid?

   b. How long is the side of the cube?
   Hint: Guess and check.

5. A conical and a cylindrical drinking glass have the same height. The top of the conical glass and the top of the cylinder are congruent circles.

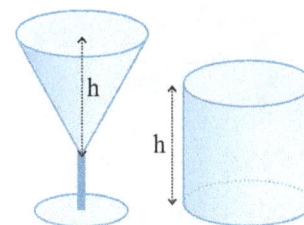

   a. What percent of the volume of the cylindrical glass is the volume of the conical glass?

   b. Assume the conical glass is full of water, and the cylindrical glass is 2/3 full. Now what percent is the volume of the water in the conical glass as compared to the volume of the water in the cylindrical glass?

6. The taller "party-hat" has a bottom diameter of 10 cm and a height of 25 cm. The diameter of the second hat is twice the diameter of the first, and its height is half the height of the first hat. Would that mean that their volumes are equal?

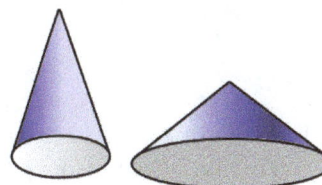

   Find out by calculating their volumes.

   What is the simple relationship between their volumes?

Volume of Spheres

A **sphere** in mathematics is a perfect ball.

The volume of a sphere is given by the formula $V = \dfrac{4}{3}\pi r^3$, where r is the radius of the sphere.

Example 1. Find the volume of a sphere with a radius of 21.0 cm.

We use the formula $V = (4/3)\pi r^3$, and get $V = (4/3)\pi(21 \text{ cm})^3$. In mathematics, it is common to give answers in terms of π, so $V = 12\,348\pi \text{ cm}^3$ is not only acceptable, but also the exact answer.

For real-life applications, we use the rounded answer. If your calculator doesn't know the order of operations, it is best to first calculate 21 cubed, and then multiply that by π and by 4, and lastly divide by 3. We get the volume is about $38\,792 \text{ cm}^3 \approx 38\,800 \text{ cm}^3$.

You may use a calculator in all problems in this lesson.

1. Find the volume of a ball with a radius of 1.5 inches, to the nearest whole cubic inch.

2. The dome of The Zeiss Planetarium in Jena, Germany, is a half-sphere, with a diameter of 23.0 metres. Calculate its volume to the nearest ten cubic metres.

3. Find out the volume of some small ball in two ways.

   a. Measure its diameter or circumference in centimetres, the best you can. Then use the formula for the volume of a sphere. Round your answer to the nearest ten cubic centimetres.

   b. Immerse the ball into a measuring cup that has some water. Record the volume of the water (in millilitres) before placing the ball in it. Then record the volume of both ball and water.

   Compare the two results, noting that one cubic centimetre equals one millilitre. If your measurements and calculations are accurate, they should be close.

> **Example 2.** A basketball is packed into a cube-shaped box with a tight fit. What percentage of the box does the ball occupy?
>
> Occupying a space means we will look at the *volume* of both the ball and the box. It seems like we would need to know the radius or the diameter of the ball to calculate this, but it turns out we don't! This is because while the radius *is* used in the formula for volume, it will get cancelled out from the fraction we use.
>
> Let r be the radius of the ball. The edge of the box is exactly twice the radius of the ball, or $2r$. The volume of the ball is $(4/3)\pi r^3$. The volume of the box is $(2r)^3 = 8r^3$. The ball therefore occupies $\dfrac{(4/3)\pi r^3}{8r^3}$ of the box. In this fraction, the r^3's cancels out, so we are left with $\dfrac{(4/3)\pi}{8} \approx 0.52360$.
>
> So, the ball only occupies <u>52.4%</u> of the box. Just a little over half of it!
>
> *Notice* that the above result is independent of the size of the ball!

4. One ball has a radius of 1 inch, and another has a radius of 2 inches.
   What fraction is the volume of the first ball of the volume of the second?
   Hint: write this fraction using the formula for the volume, and simplify it.

5. Three tennis balls fit snugly in a cylindrical container. Calculate what fraction of the total volume of the cylinder the tennis balls take up.
   Hint: write this fraction using the formulas for the volumes, and simplify it.

6.0 cm

6. The earth has a shape that is close to a sphere. In reality, it is slightly "flattened" at the poles, but in this exercise we will treat it as a perfect sphere, with a (mean) diameter of 12 740 km. Calculate its volume. Round your answer to the nearest billion cubic kilometres.

   <u>Note:</u> *You may need a computer or an online calculator to handle such a large number.*

7. Find the total mass of 50 glass marbles, to the nearest ten grams. Assume that the marbles are spheres with a diameter of 16 mm. The density of glass is 2.6 g/cm³, which means that each cubic centimetre of glass has a mass of 2.6 grams.

Volume Problems

You may use a calculator in all problems in this lesson.

1. Memorize the above formulas.

   Hint: notice how the formula for the volume of a sphere is similar to the formula for the area of a circle. The area of a circle is "pi *r* squared." The formula for volume has "pi *r* <u>cubed</u>", and then it is multiplied by the fraction 4/3.

2. A large tank consisting of a cylindrical top and a conical bottom (for easy drainage) is being filled with biodiesel at a rate of 56.6 litres per minute. How long will it take to fill it?

   The diagram below will help you figure out how many cubic centimetres are in one cubic metre. Note also that 1 m³ = 1000 L.

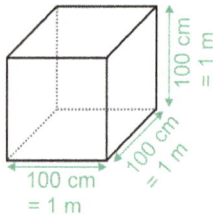

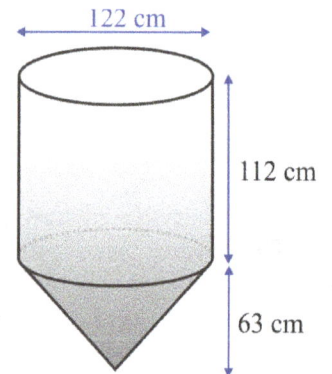

3. Many objects are in the shape of a **frustum**, or a cut cone. It is like a cone from which a smaller cone is cut off. There exists a formula for its volume, but you can calculate the volume without it, if you know the dimensions of the "cut" part. (Think subtraction.)

   a. Find the volume of the frustum on the right, to the nearest hundred cubic centimetres.

   b. Convert that to litres (1 L = 1000 cm³).

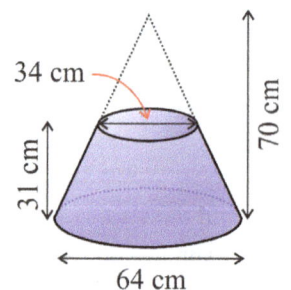

4. Elkhart Ammonium Nitrate Storage in Elkhart, Texas, is a large building consisting of a half sphere on top of a circular cylinder. The stem wall is 11.0 m high, and the diameter of the circle (which is also the diameter of the sphere) is 17.5 m. Calculate its volume to the nearest hundred cubic metres.

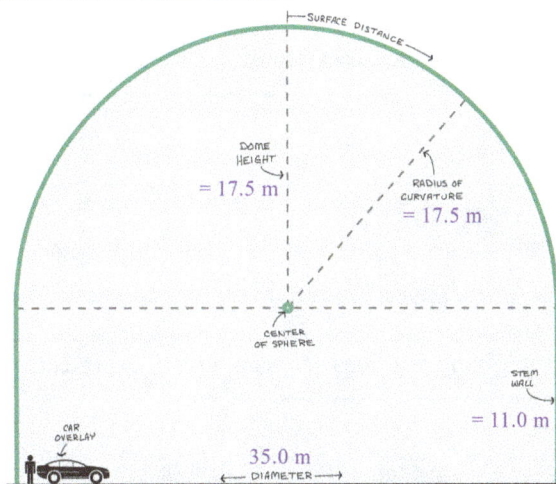

Images courtesy of Monolithic Dome Institute, www.monolithic.com

5. A company that makes specialty fruit jams uses jars in the shape of a hexagonal prism. This not only looks fancier than a regular, cylindrical jar, but has a smaller volume than a cylindrical jar with the same height. This way the company can use less jam to fill the jar, yet charge the same amount, or a bit more, than competitors who use cylindrical jars. To the consumer, both types of jars tend to look like they contain the same amount, if they have the same height.

On the right, you see the bottom face of the jar. We can divide it into six congruent equilateral triangles, each with a height of 2.8 cm.

a. Assume the jar is 8 cm tall. Calculate its volume to the nearest cubic centimetre.

b. Calculate the volume of a (circular) cylindrical jar with the same height, and a diameter of 6.4 cm, to the nearest cubic centimetre.

c. What percentage is the volume of the smaller jar of the volume of the bigger?

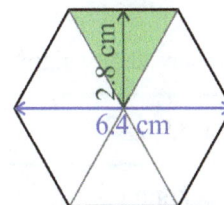

Mixed Review Chapter 2

1. Find the value of the expressions.

a. $-4^2 =$	**b.** $5^{-3} =$	**c.** $57 \cdot 10^{-3} =$	**d.** $3^{-6} \cdot 3^9 \cdot 3^{-2} =$
e. $\dfrac{8^6}{8^8} =$	**f.** $\left(\dfrac{-1}{-5}\right)^2 =$	**g.** $2^{-2} \cdot 5 \cdot 2^5 + 4^3 \cdot 4^{-2} =$	

2. Simplify, giving your answer without any exponents.

a. $4 \cdot 10^2 \, 5^2 =$	**b.** $2 \cdot 7^{-1} 7^{-5} 7^3 =$	**c.** $6^9 \cdot 6^{-9} =$

3. Write an equivalent expression using the exponent laws, without using negative exponents.

a. $(2b^5)^3 =$	**b.** $(3y)^{-3} =$	**c.** $6x^{-2}$	**d.** $-2a^3 b^7 \cdot 5a^7 b^2 =$
e. $\dfrac{x^{-3}}{x^{-2}} =$	**f.** $\dfrac{2a^2}{a^5} =$	**g.** $\left(\dfrac{-2x}{3}\right)^3 =$	**h.** $\left(\dfrac{x}{z^2}\right)^2 =$

4. Compare the numbers, writing $<$ or $>$ in the box.

    a. $6 \cdot 10^{-6}$ ☐ $8 \cdot 10^{-7}$ **b.** $5 \cdot 10^{-5}$ ☐ 0.0004 **c.** $2.9 \cdot 10^{-4}$ ☐ 0.0003

5. Anna says that $(5x)^{-2}$ simplifies to $5x^{-2}$. Is she correct? If not, why not?

6. How many significant digits do these numbers or quantities have?

a. 15.0	**b.** 150	**c.** 0.15	**d.** 0.150
e. 4000 kg	**f.** 4001 kg	**g.** $7.02 \cdot 10^6$	**h.** $2.060 \cdot 10^{-3}$

7. Rewrite the numbers in scientific notation correctly.

    a. $403 \cdot 10^{-3}$ **b.** $66 \cdot 10^6$

    c. $0.2 \cdot 10^{-2}$ **d.** $0.291 \cdot 10^{-1}$

8. Simplify. Give your answers in decimal notation (not scientific).

a. $2 \cdot 10^{-3} + 5 \cdot 10^{-2}$	**b.** $7 \cdot 10^{-5} + 0.03$
c. $3.2 \cdot 10^{-1} - 0.07$	**d.** $5.4 \cdot 10^4 - 2000 + 8 \cdot 10^3$

9. Calculate with a calculator. Round your answer to the correct amount of significant digits.

a. 0.92 m · 1.3 m	**b.** 121 L ÷ 1302 km
c. 39 800 km ÷ 3.14	**d.** 6.23 kg · 4200
e. $(4.50 \cdot 10^6$ dollars) ÷ 13 000 people	**f.** 18 · 35 kg · 115

10. The surface area of the sun is $6.09 \cdot 10^{12}$ km^2, and that is $12 \cdot 10^3$ times the surface area of the earth. Find the surface area of the earth. Use scientific notation.

11. Calculate. Give each answer to a reasonable accuracy.

   a. A certain city spent 977.9 million dollars for education in a particular year, and they have 550 000 students (accurate to thousands). How much did they spend per student?

   b. What is the area of a triangle with a 56-cm base and 34-cm height?

12. If the distance you need to travel is $9.3 \cdot 10^7$ mi (from the earth to the sun), and the time you take to do that is $1.38 \cdot 10^{-1}$ hours, then what is your speed, in miles per hour? (Give your answer in scientific notation.)

Chapter 2 Review

1. Describe a sequence of transformations that can map figure 1 to figure 2.

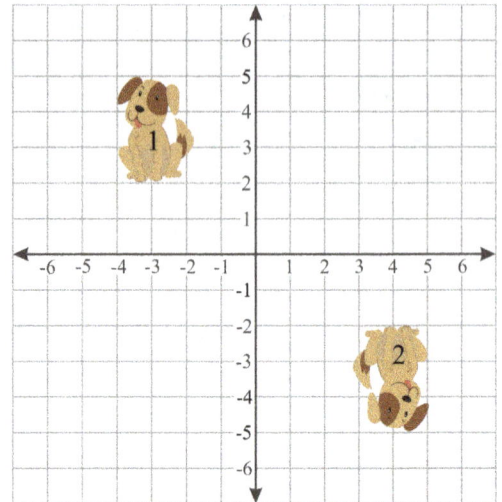

2. Rotate angle DEF both 90 degrees and also 180 degrees clockwise around the origin.

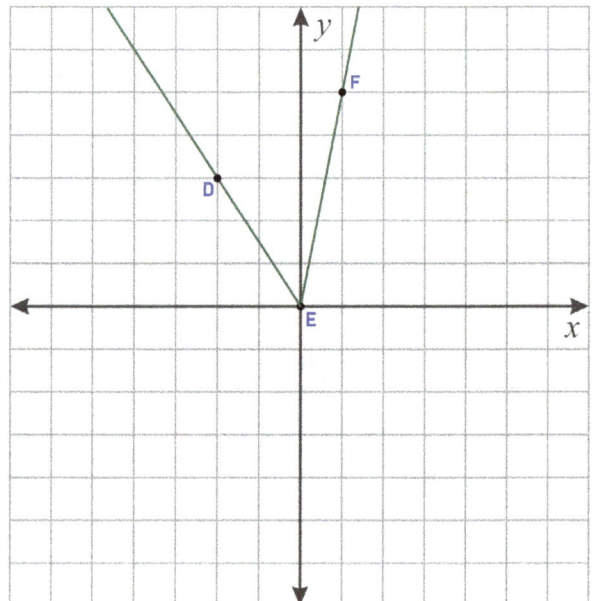

3. A quadrilateral was first reflected in the *y*-axis, and then rotated around the origin clockwise 90 degrees. Its vertices are now at points (3, −5), (5, −2), (4, −1), and (1, −4). What were the coordinates of its vertices before these transformations?

4. Figure A'B'C'D'E' is a dilation of figure ABCDE with scale factor 3/4. Angle B is a right angle. Check all the statements that are true.

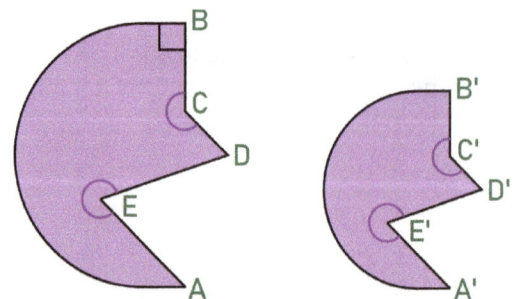

   a. Angle B' is a right angle.

   b. The measure of ∠CDE is 3/4 of the measure of ∠C'D'E'.

   c. ∠E is equal to ∠E'.

   d. If CD = 1 inch, then C'D' = 3/4 inch.

   e. ∠D is equal to ∠E'.

   f. If the perimeter of figure ABCDE is 20 inches, then the perimeter of A'B'C'D'E is 12 inches.

5. **a.** Perform the following sequence of transformations to triangle ABC:

First rotate it counterclockwise around point C 90 degrees.
Then reflect it in the vertical line at $x = -1$.
Lastly, translate it two units to the right and three down.

b. Find another, different sequence of transformations that does the same as the sequence in (a), and starts with a reflection.

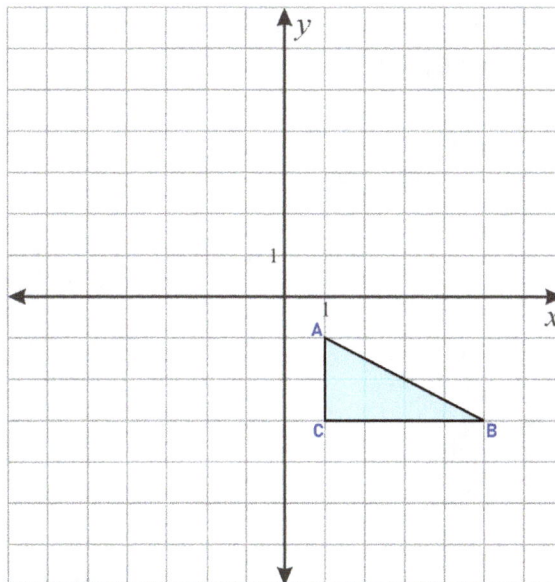

6. Draw a dilation of triangle ABC...

a. from point C and scale factor 1/2.	**b.** from point B and scale factor 2.
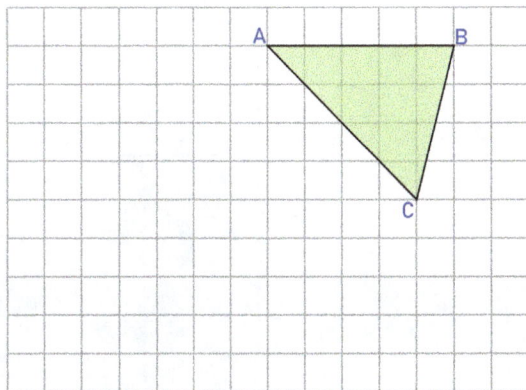	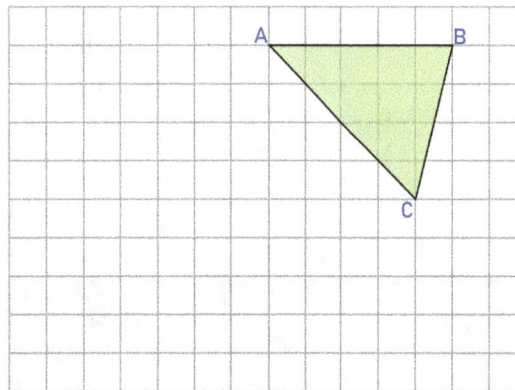

7. Parallelogram EFGH underwent the following transformations :

1. Reflection in the vertical line at $x = -0.5$.

2. Translation 3 units to the left and 4 units down.

3. Dilation from point E" with scale factor 2.

What are the coordinates of the image of point F after all these transformations?

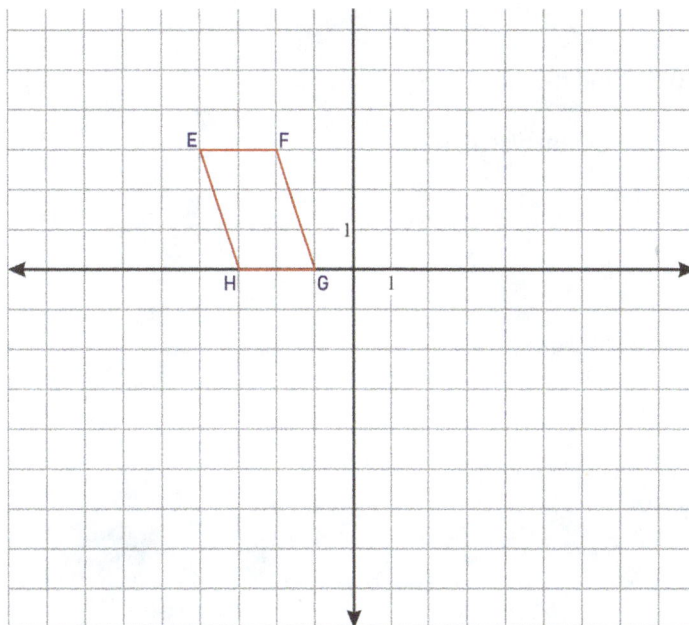

8. Show that the two triangles are similar by describing a sequence of transformations that could map △DEF to the smaller triangle

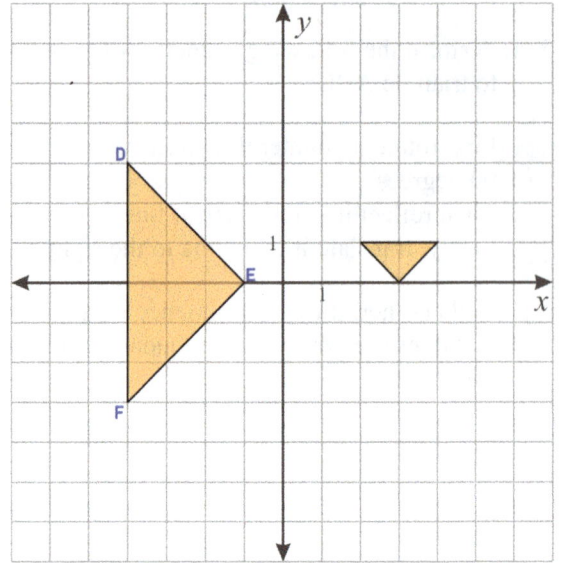

9. Figure PQRS underwent a dilation, then a rotation. Study the coordinates to find out the details about each transformation, then fill in the missing coordinates.

Original figure	Dilation	Rotation
P(−5, 3)	P'(−6, 5)	P"(___ , ___)
Q(0, 3)	Q'(4, 5)	Q"(___ , ___)
R(−1, 1)	R'(___ , ___)	R"(−4 , −5)
S(−4, 1)	S'(−4, 1)	S"(−4, 1)

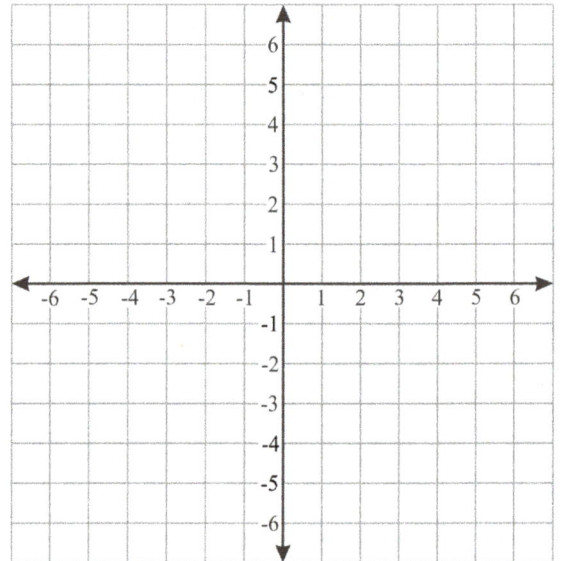

10. **a.** Find the value of x.

   b. Find the value of α.

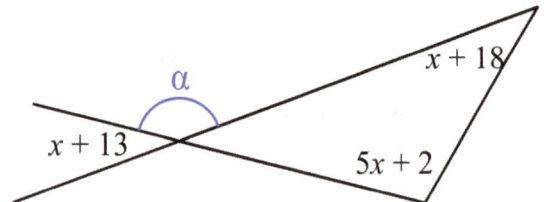

11. Lines *l* and *m* are parallel. Figure out the measure of the angle α. (You may need to mark more angles in the diagram.)

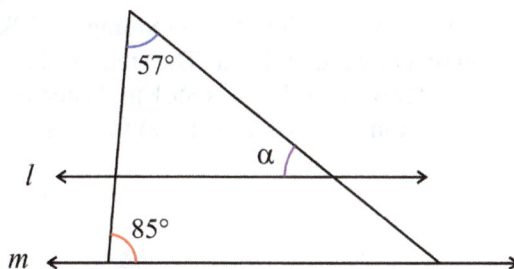

12. Lines *m* and *n* are parallel. Find the measure of angle *x*, and prove why it is what you find it to be. In other words, explain and justify your reasoning. You may need to mark more angles in the diagram.

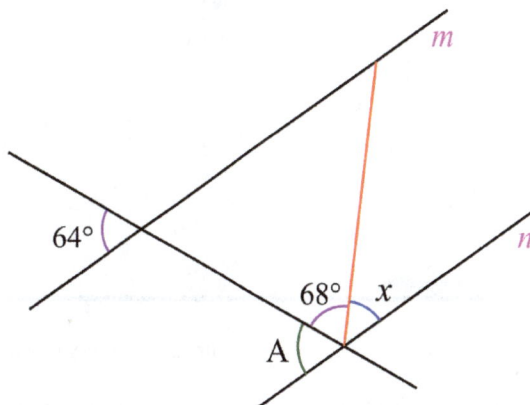

13. A shampoo bottle is in the shape of a circular cylinder. It says it contains 473 ml of shampoo. Its inner diameter is 6.0 cm and its height is 17 cm. What percent of the bottle does the shampoo take up?

14. Compare a sphere with radius 5 cm with a cone with the same radius. What is the height of the cone, given the two have the same volume?

15. Elkhart Ammonium Nitrate Storage in Elkhart, Texas, is a large building consisting of a half sphere on top of a circular cylinder. The stem wall is 11.0 m high, and the diameter of the cylinder (which is also the diameter of the sphere) is 35.1 m. When the storage is filled with ammonium nitrate, the top part of it (the part inside the half-sphere) forms a cone.

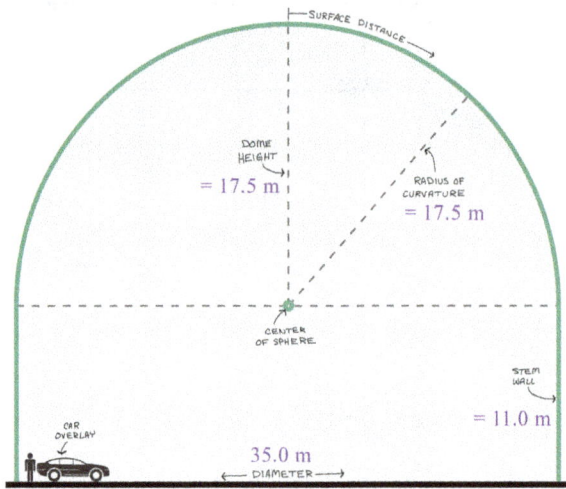

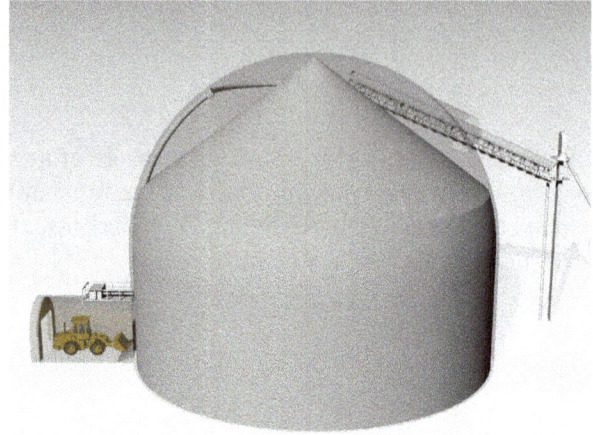

Images courtesy of Monolithic Dome Institute, www.monolithic.com

Find the volume of the ammonium nitrate mound when the cone reaches the top of the structure, to the nearest hundred cubic meters.

Chapter 3: Linear Equations
Introduction

The third chapter of Math Mammoth Grade 8 focuses both on the mechanics of solving linear equations and on problem solving.

The chapter starts with a vocabulary reference sheet. The first actual lesson is a review of integer addition and subtraction, which you can omit at your discretion. The next several lessons after that review simple equations of the form $px + q = r$ and $p(x + q) = r$ and the distributive property from 7th grade.

The next step towards solving more complex equations is the lesson *Combining Like Terms*. Students add and subtract like terms, including with decimal or fractional coefficients, and solve equations where like terms need combined first.

Having learned this, students then tackle some typical algebraic word problems in the following lesson.

Then it is time to learn to solve equations where the variable is on both sides. There are often several possible solution pathways. Students also learn about the common error of adding or subtracting "across the sides."

The lesson *Simplifying Linear Expressions* focuses on how to remove brackets after a minus sign, such as in the expression $2(3 + 2y) - 7(3 - 5y)$. After that, it is time for more practice and word problems, including age and coin word problems.

Then we turn our attention to equations with fractions, and the student learns to multiply both sides of the equation by a common multiple of the denominators. In the lessons on formulas, the student both solves various formulas for a variable in it, and uses formulas to solve a variety of word problems.

The lesson *More on Equations* deals with equations that have an infinite number of solutions (identities) or no solutions.

The chapter ends with two more lessons on word problems (percent word problems and miscellaneous problems).

Pacing Suggestion for Chapter 3

This table does not include the chapter test as it is found in a different book (or file).
Please add one day to the pacing if you use the test.

The Lessons in Chapter 3	page	span	suggested pacing	your pacing
Algebra Terms (For Reference)	125	*(1 page)*		
Review: Integer Addition and Subtraction	126	*3 pages*	1 day	
Equations Review, Part 1	129	*4 pages*	1 day	
The Distributive Property	133	*3 pages*	1 day	
Equations Review, Part 2	136	*4 pages*	1 day	
Equations Review, Part 3	140	*4 pages*	1 day	
Combining Like Terms	144	*3 pages*	1 day	
Word Problems	147	*4 pages*	1 day	
A Variable on Both Sides	151	*4 pages*	1 day	

The Lessons in Chapter 3	page	span	suggested pacing	your pacing
Word Problems and More Practice	155	*3 pages*	1 day	
Simplifying Linear Expressions	158	*3 pages*	1 day	
More Practice	161	*3 pages*	1 day	
Age and Coin Word Problems	164	*3 pages*	1 day	
Equations with Fractions 1	167	*3 pages*	1 day	
Equations with Fractions 2	170	*3 pages*	1 day	
Formulas, Part 1	173	*2 pages*	1 day	
Formulas, Part 2	175	*2 pages*	1 day	
More on Equations	177	*3 pages*	1 day	
Percent Word Problems	180	*2 pages*	1 day	
Miscellaneous Problems	182	*2 pages*	1 day	
Chapter 3 Mixed Review	184	*2 pages*	1 day	
Chapter 3 Review	186	*3 pages*	1 day	
Chapter 3 Test (optional)				
TOTALS		*63 pages*	21 days	

Helpful Resources on the Internet

We have compiled a list of Internet resources that match the topics in this chapter, including pages that offer:

- **online practice** for concepts;
- online **games**, or occasionally, printable games;
- **animations** and interactive **illustrations** of math concepts;
- **articles** that teach a math concept.

We heartily recommend you take a look! Many of our customers love using these resources to supplement the bookwork. You can use these resources as you see fit for extra practice, to illustrate a concept better and even just for some fun. Enjoy!

https://l.mathmammoth.com/gr8ch3

Scan me

Algebra Terms
For Reference

Expressions in mathematics consist of: • numbers; • mathematical operations ($+$, $-$, $\cdot$, $\div$, exponents); • and letter variables, such as x, y, a, T, and so on. Note: Expressions do *not* have an equals sign!	**Examples of expressions:** $\frac{3}{5}x^2 - 3x + 5$ $\quad$ 5 $\quad$ $\left(\dfrac{3x}{y^2}\right)^2$ $T - 29$ $\qquad$ $2^x - 5^y$
An **equation** has two expressions separated by an equals sign: <div align="center">(expression 1) **=** (expression 2)</div>	**Examples of equations:** $0 = 0$ $\qquad$ $2(z - 9) = -z^2$ $9 = -8$ $\qquad$ $\dfrac{x + 3}{2} = -1.5$ (a false equation)

A **term** is an expression that consists of numbers and/or variables that are *multiplied*. For example, $7x$ is a term and so is $0.6mn^2$.

A single number or a single variable is also a term. If the term is a single number, such as 4.5 or ¾, we call it a **constant**.

In the expression on the right, we have three terms: $5xy^2$, $\frac{2}{3}x$, and 9, that are separated by subtraction and addition.

If a term is not a single number, then it has a **variable part** and a **coefficient**.

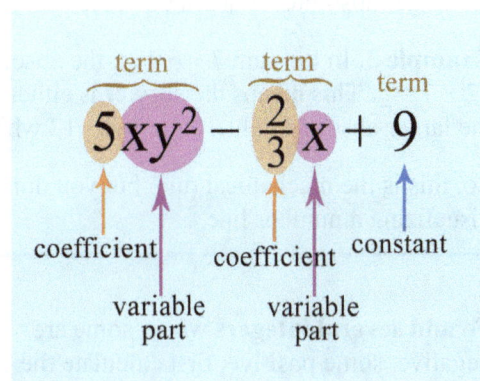

$$5xy^2 - \frac{2}{3}x + 9$$

term · term · term
coefficient · coefficient · constant
variable part · variable part

• The coefficient is the single number by which the variable or variables are multiplied.

• The variable part consists of the variables and their exponents.

For example, in $4.3ab$, 4.3 is the coefficient, and ab is the variable part.

Note: a term that consists of variables only still has a coefficient: it is one. For example, the coefficient of the term x^3 is one, because you can write x^3 as $1 \cdot x^3$.

Example. Is $s - 5$ a term? No, it is not since it contains subtraction. Instead, $s - 5$ is an expression consisting of two terms, s and 5, separated by subtraction.

1. Write the expression based on the clues.

• It has four terms.
• The constant term is the square of the third smallest prime.
• The variable parts of the variable terms are ab, a^2, and a, respectively.
• The coefficients of the variable terms are the three consecutive integers with a sum of 21.
• The first two terms are separated by subtraction, the rest by addition.

Review: Integer Addition and Subtraction

Integers consist of the counting numbers (1, 2, 3, 4, ...), zero, and the negative counterparts of the counting numbers (−1, −2, −3, −4, ...). So, the set of integers is {..., −4, −3, −2, −1, 0, 1, 2, 3, 4, ...}.

An **absolute value** of an integer is its distance from zero, and is marked with two vertical lines. For example, $|2| = 2$ and $|-18| = 18$.

We obtain the **opposite** or **negation** of an integer by changing its sign from positive to negative, or vice versa. For example, the opposite or negation of 17 is −17. The opposite of −4 is 4.
We can use the negative sign "−" to signify this: −(−5) means the opposite of −5, which is 5.

To **add several negative integers**, simply add their absolute values and write the answer as negative.	**Example 1.** To find the sum $-8 + (-3) + (-7) + (-11)$, add $8 + 3 + 7 + 11 = 29$. The value of the original sum is −29.

To **add a negative and a positive integer**, find the difference in their absolute values. The integer with the bigger absolute value determines the sign of the final answer.

Example 2. In the sum $-9 + 11$, the absolute values of the two integers are 9 and 11. Their difference is $11 - 9 = \underline{2}$. This means the answer is either 2 or −2. To determine which, check the sign of the integer with the larger absolute value. In our case it is 11 (which is positive), so the answer is 2 (and not −2).

Example 3. In the sum $7 + (-12)$, the absolute values of the two integers are 7 and 12. Their difference is $12 - 7 = \underline{5}$. This means the answer is either 5 or −5. To determine which, check the sign of the integer with the larger absolute value. Here it is −12 which is negative, so the answer is −5 (and not 5).

So, this is the mechanical rule, but you don't have to use it if you have learned other methods, such as visualizing a number line.

To **add several integers** where some are negative, some positive, first calculate the partial sums of all the negative integers and of all the positive ones. Lastly add those sums.	**Example 4.** $-8 + 12 + (-9) + (-1) + 5 + (-6) = ?$ Positives: $12 + 5 = 17$ Negatives: $-8 + (-9) + (-1) + (-6) = -24$ Total: $17 + (-24) = \underline{-7}$

1. Add.

a. $(-4) + 8 =$	**b.** $15 + (-25) =$	**c.** $-12 + 6 =$	**d.** $-11 + (-32) =$
e. $-12 + (-2) + (-5) =$		**f.** $6 + (-1) + (-5) + 2 =$	**g.** $-7 + 10 + (-6) + 1 =$
h. $-11 + (-2) + 7 + (-5) + 4 + (-3) =$		**i.** $-6 + (-5) + 8 + (-12) + 24 + 1 =$	

To **subtract two integers**, you can often think with the help of the number line model. For example, you can visualize $2 - 6$ as starting at 2, and moving 6 steps to the left on the number line.

Mathematicians actually define the subtraction of two numbers, $a - b$, as the <u>sum</u> of a and the opposite of b.

In symbols: $a - b = a + (-b)$

In other words, to subtract an integer, change the subtraction to an addition of the opposite number. From this definition it also follows that $a - (-b)$ simplifies to $a + b$.

(Why? In $a - (-b)$ we subtract $-b$, and the opposite of $-b$ is b. Instead of subtracting $(-b)$, you add *its opposite*, or b.)

Example 5.

$5 - 7$	$-6 - 8$	$2 - (-4)$	$-3 - (-9)$
↓	↓	↓	↓
$5 + (-7) = -2$	$-6 + (-8) = -14$	$2 + 4 = 6$	$-3 + 9 = 6$

2. Write each subtraction as an addition, and solve.

a. $8 - (-6)$	**b.** $-9 - (-14)$	**c.** $-21 - 8$	**d.** $3 - 15$
↓	↓	↓	↓
$8 + 6 = $ _____	___ $+$ ___ $= $ _____	___ $+$ ___ $= $ _____	___ $+$ ___ $= $ _____

3. Subtract.

a. $2 - 9 =$	**b.** $-2 - 9 =$	**c.** $-2 - (-9) =$	**d.** $2 - (-9) =$
e. $-7 - 4 =$	**f.** $-7 - (-4) =$	**g.** $7 - (-4) =$	**h.** $4 - 7 =$

4. Can any addition be changed to a subtraction? See if you can find matching subtractions for these additions.

a. $7 + (-10)$	**b.** $-2 + (-1)$	**c.** $14 + 3$	**d.** $-8 + 5$
___ $-$ ___ $= $ _____	___ $-$ ___ $= $ _____	___ $-$ ___ $= $ _____	___ $-$ ___ $= $ _____

So, any subtraction can be written as an addition. The converse is also true: any addition can be written as a subtraction. For example, the sum $5 + 4$ can be written as $5 - (-4)$, and the sum $2 + (-13)$ can be written as the subtraction $2 - 13$.

In symbols, $c + d = c - (-d)$. Instead of adding d, you subtract the opposite (or negation) of d.

However, since this usually does not simplify the calculation, it does not get used often.

5. Solve, working in order from left to right.

a. $-2 - 5 + 6 =$	**b.** $7 + (-12) - 5 - 8 =$	**c.** $-1 + 9 - 14 + 7 =$
d. $-8 - 12 - 5 + 9 =$	**e.** $2 - (-12) - 10 - (-3) =$	**f.** $-21 - 13 + 8 - (-5) =$

Adding negative fractions

Example 6. Add $\dfrac{2}{5} + \left(-\dfrac{3}{7}\right)$. We could use the regular

process for integer addition: figure out which fraction has a larger absolute value, then subtract the smaller absolute value from the larger one, and so on. But the easier way is this: Simply add

the fractions normally and treat the negative fraction $-\dfrac{3}{7}$ as $\dfrac{-3}{7}$.

You will end up with an *integer addition* in the numerator. See the full solution on the right.

$$\dfrac{2}{5} + \left(-\dfrac{3}{7}\right)$$

$$\dfrac{2}{5} + \dfrac{-3}{7}$$

$$\dfrac{14}{35} + \dfrac{-15}{35}$$

$$\dfrac{14 + (-15)}{35} = \dfrac{-1}{35} = -\dfrac{1}{35}$$

6. Add and subtract.

a. $\dfrac{2}{7} + \left(-\dfrac{3}{4}\right)$

b. $\dfrac{1}{8} + \left(-\dfrac{1}{2}\right) + \left(-\dfrac{3}{4}\right)$

c. $-\dfrac{5}{6} + \dfrac{2}{9}$

d. $-\dfrac{2}{3} + \dfrac{2}{9} - \dfrac{1}{6}$

e. $-\dfrac{7}{8} + \left(-\dfrac{1}{10}\right)$

f. $\dfrac{1}{6} - \left(-\dfrac{7}{8}\right)$

Equations Review, Part 1

An **equation** consists of two expressions, separated by an equals sign:

> **expression 1** = **expression 2**

For example, $40 = w + 32$ is an equation, and so is $2 = 5$, the latter being a *false* equation.

A **solution** or a **root** to an equation is a value of the unknown that makes the equation *true*; in other words, makes the two expressions on both sides to have the same value.

Example 1. Is 20 a root to the equation $11 = \frac{1}{2}x + 3$?

To check that, we substitute 20 in place of x and check whether the two sides of the equation have the same value:

$$11 \stackrel{?}{=} \frac{1}{2}(20) + 3$$

$$11 \neq 10 + 3$$

No, 20 does not fulfill this equation, so it is not a root.

1. **a.** Is 2 a root to the equation $\frac{3x^2 - 7}{5} = x$? Explain.

   b. Is -90 a root to the equation $\frac{2}{3}y + 11 = -49$? Explain.

2. Without solving the equation, check whether $x = -3$ is a solution to the equation $x + 4x + 6x - 8 = -5(x + 8)$. Before you start, think: would you be allowed to simplify the left side of the equation?

3. Write three different equations with the solution of $x = -5$.

4. If $2w + 6 = 50$ and $3w - 15 = 51$, then does $2w + 6$ equal $3w - 15$? Justify your reasoning.

To solve an equation, we perform the same mathematical operation (add, subtract, multiply, divide) to *both* sides of the equation. Notice that in that process, <u>the two sides of the equation remain equal</u>, even though the expressions themselves, on both sides, change!

<u>Note:</u> We can **mark the operation to be done to both sides** either below each line of the solution or in the right margin, after a vertical line. I prefer marking it in the right margin, because that is how I was taught in school in Finland, but you can go with whatever you or your teacher prefers.

Example 2.	**Example 3.**	**Example 4.**
$7w + 8 = 43$ $\underline{\quad -8 \quad -8 \quad}$ $7w = 35$ $\dfrac{7w}{7} = \dfrac{35}{7}$ $w = 5$	$-14 = \dfrac{1}{5}a$ $5(-14) = 5 \cdot \dfrac{1}{5}a$ $-70 = a$ $a = -70$	$-18 - x = 24 + 8$ $-18 - x = 32 \quad\Big|\; + 18$ $-x = 50 \quad\Big|\; \div (-1)$ $x = -50$
Check: $7(5) + 8 \overset{?}{=} 43$ $43 = 43$ ✓	Check: $-14 \overset{?}{=} \dfrac{1}{5}(-70)$ $-14 = -14$ ✓	Check: $-18 - (-50) \overset{?}{=} 24 + 8$ $-18 + 50 \overset{?}{=} 32$ $32 = 32$ ✓

5. See Derek's solution on the right.

   a. Check whether $x = 14$ is truly a root.

   b. If not, correct the error in his solution.

$$10 - x = 24$$
$$\underline{\; - 10 \quad - 10 \;}$$
$$x = 14$$

6. Solve the equations. Check your solutions.

a. $\quad -p = 12 - 34$	**b.** $\quad 78 - x = -8$	**c.** $\quad -2 - 7 = -3z$
d. $\quad \dfrac{y}{-4} = -22$	**e.** $\quad 10x = -40 - 5$	**f.** $\quad 2.1 - x = -6.7$

Remember, our goal is to **isolate the unknown** (have it alone on one side).

This example shows a typical two-step equation. On the side of the unknown (left), there is both a multiplication by 8 and an addition of 7. We need to undo both of those operations, in two steps.

Example 5.

$$8x + 7 = -5 \qquad \Big|\, -7$$
$$8x = -12 \qquad \Big|\, \div 8$$
$$x = -12/8$$
$$x = -3/2$$

Check:

$$8 \cdot (-3/2) + 7 \overset{?}{=} -5$$
$$-24/2 + 7 \overset{?}{=} -5$$
$$-12 + 7 \overset{?}{=} -5$$
$$-5 = -5 \quad \checkmark$$

Note: In algebra, it is preferable to leave the answer as a fraction, not as a mixed number. This is to avoid confusion, because $-1\ 1/2$ could be confused with $-11/2$ if not written clearly. In some real-life contexts a mixed number may be preferable.

7. Solve the equations. Check your solutions. (Give your answer as a fraction, not mixed number, when applicable.)

a. $\quad 5x + 2 = 67$	**b.** $\quad -3y + 2 = 71$	**c.** $\quad 25 - 3w = 17$
d. $\quad -34 = 2x - 11$	**e.** $\quad -98 = -8z - 2$	**f.** $\quad -8 - 4z = 10$

8. The solution on the right shows a common student error. We can verify
the root is *not* ½ by substituting it to the original equation:

$$14 - 80(½) \stackrel{?}{=} 54$$

$$14 - 40 \stackrel{?}{=} 54$$

$$-26 \neq 54$$

What is the error? Correct it.

$$14 - 80x = 54$$
$$\underline{-14 \quad\quad -14}$$
$$80x = 40$$
$$\frac{80x}{80} = \frac{40}{80}$$
$$x = ½$$

9. Use these for more practice, as necessary. (Give your answer as a fraction, not mixed number, when applicable.)

a. $\quad -5 + 15 = -6w$	**b.** $\quad 6 = \dfrac{d}{-1.1}$	**c.** $\quad \dfrac{a}{5} = -1.2 + (-3.1)$
d. $\quad 56 - 5x = 28$	**e.** $\quad -35 = -4q + 2$	**f.** $\quad -150 + 30w = 60$
g. $\quad 13.5 - 2y = 7$	**h.** $\quad 7.8 - 16.2 = \dfrac{x}{7}$	**i.** $\quad -55 = -6w - 13$

The Distributive Property

The **distributive property** states that we can distribute multiplication over addition: $a(b + c) = ab + ac$.

It also applies with subtraction: $a(b - c) = ab - ac$.

Example 1. Each term needs to be multiplied by the factor in front of the brackets. In other words, below, you need to "take the 5 through" onto *everything* inside the brackets:

$$5(x + y + 6) = 5x + 5y + 5 \cdot 6, \text{ which simplifies to } 5x + 5y + 30$$

Example 2. Be careful with negative numbers and with subtraction. Compare the two examples on the right.

In both, we take -2 through the brackets. Notice how the "+" sign in $x + 1$ and the "−" sign in $x - 1$ is preserved.

(For clarity, I've written the equivalent expressions one under the other, instead of as one long horizontal line.)

$$-2(x + 1)$$
$$-2 \cdot x + (-2) \cdot 1$$
$$-2x + (-2)$$
$$-2x - 2$$

$$-2(x - 1)$$
$$-2 \cdot x - (-2) \cdot 1$$
$$-2x - (-2)$$
$$-2x + 2$$

1. The two multiplications on the right show a common student error. What is that error?

$$4(a + b + 3)$$
$$\downarrow$$
$$4a + 4b + 3$$

and

$$7(2x - 9)$$
$$\downarrow$$
$$14x - 9$$

2. Multiply using the distributive property. Write your answer below the original. Compare the problems. Be careful with negative numbers, and be on the lookout for a shortcut.

a. $-2(x + 9)$	**b.** $-2(x - 9)$	**c.** $-3(5x + 8)$	**d.** $-3(5x - 8)$	**e.** $-5(2x + 7)$	**f.** $-5(2x - 7)$

Example 3. Study these examples of a quicker way of removing the bracket.

If you're unsure of this, you can use the longer route as shown in example 2.

$$-5 (2x - 6)$$
$$-5 (2x) - 5 (-6)$$
$$-10x + 30$$

$$-7 (-3x + 6y - 4)$$
$$-7 (-3x) - 7 (6y) - 7 (-4)$$
$$21x - 42y + 28$$

3. Multiply using the distributive property. Compare the problems.

a. $-2(x + y - 9)$	**b.** $-2(x - y + 9)$	**c.** $-3(9 + 2y)$	**d.** $-3(9 - 2y)$

Example 4. To simplify the expression $-(x + 8)$, think of it as $-1(x + 8)$ (*Why?). We can now use the distributive property →

$$-1(x + 8)$$
$$-1(x) + (-1)(8)$$
$$-x - 8$$

In a nutshell, $-(x + 8) = -x - 8$. It is as if we take the minus sign through the brackets, and distribute it to each individual term. This *changes the sign* of each individual term inside the brackets.

In yet other words, $-(x + 8)$ is the opposite of the expression $x + 8$, and it is obtained by changing each individual term to its opposite.

4. Find the opposites of the expressions.

a. $-(y + 3)$	**b.** $-(y - 3)$	**c.** $-(6 - a + 2b)$	**d.** $-(-6 + a - 2b)$

5. Multiply using the distributive property.

a. $-0.9(a - 2b + 7)$	**b.** $-(-x + 4 - 5y)$	**c.** $-(-7w + 0.5)$	**d.** $20(-0.4 + 0.5x)$
e. $\frac{1}{4}(g - 2h + 4)$	**f.** $32(v + \frac{1}{4}w - \frac{1}{8})$	**g.** $-\frac{2}{5}(10p - 20q)$	**h.** $-100(-0.01 + 0.1z)$

Example 5. The expression $(8 + 2x)(7)$ means the same as $(8 + 2x) \cdot 7$, and it is equivalent to $7(8 + 2x)$.

Why? Because multiplication is commutative (can be done in any order). Here, $8 + 2x$ and 7 are the two factors being multiplied, and it doesn't matter in which order we multiply them.

6. Simplify.

a. $(8 + 2x) \cdot 7$	**b.** $(2y - 8 - 3x)(5)$	**c.** $(0.6x - 1.4)(-10)$
d. $(60s - 12 + 39t) \cdot \frac{1}{3}$	**e.** $(12w - 4) \cdot 6 + 15w$	**f.** $-\frac{3}{5}(25x - 45y) + 6x$

*Negative one times a number is equal to the opposite of the number: $-1 \cdot a = -a$. Considering $(x - 6)$ as a *single* number, we get $-1(x - 6) = -(x - 6)$.

When we use the distributive property "backwards," and write an expression *as a product*, it is called **factoring**. For example, $-25x + 5$ can be written as $-5(5x - 1)$.

7. Factor the expressions (write them as a product). Often, there are several ways to do it.

a. $14x - 10$	**b.** $-27x + 36$	**c.** $-12s + 18t - 6$	**d.** $-0.9x + 1.2y - 0.6$
e. $\dfrac{1}{4}w - \dfrac{1}{2}$	**f.** $\dfrac{3}{4}s + \dfrac{3}{8}$	**g.** $\dfrac{1}{5}y - x + \dfrac{6}{5}$	**h.** $-x - \dfrac{2}{3}y + \dfrac{1}{3}$

8. Find the missing term or factor.

a. $-2(6 + \boxed{}) = -12 - 18x$	**b.** $\boxed{}(4y - 5) = -12y + 15$	**c.** $\boxed{}(4v - 6w + 5) = -0.8v + 1.2w - 1$

9. Ruth thinks that $-(-3x - 6)$ simplifies to $3x - 6$, whereas Jonathan feels it simplifies to $3x + 6$. Who is right? Why?

10. Which of these expressions are equivalent?

   a. $-7(x + 1)$ **b.** $7 - 7x$ **c.** $-7x + 1$ **d.** $-1 - 7x$ **e.** $(-1 - x) \cdot 7$

Division, too, can be distributed over addition or subtraction:

$$\frac{a + b}{c} = \frac{a}{c} + \frac{b}{c} \quad \text{and} \quad \frac{a - b}{c} = \frac{a}{c} - \frac{b}{c}$$

---

11. Fill in the missing portions.

a. $\dfrac{8y + 48}{8} = \dfrac{\boxed{}}{8} + \dfrac{\boxed{}}{8} = \boxed{} + 6$	**b.** $\dfrac{2x - 12}{6} = \dfrac{\boxed{}}{6} - \dfrac{12}{6} = \dfrac{\boxed{}}{3} - \boxed{}$

12. Simplify.

a. $\dfrac{25x + 30y}{5} =$	**b.** $\dfrac{1.5 - 2.4a + 9b}{3} =$	**c.** $\dfrac{16x - 2y + 24}{8} =$

Puzzle Corner   Simplify.   **a.** $\dfrac{5x^2 - 2x}{x} =$ **b.** $\dfrac{-30y^2 + 12y}{6y} =$

Equations Review, Part 2

Sometimes you have to use the distributive property to remove the brackets. See example 1 on the right. Checking the solution, we get:

$$2(-14 + 9) \overset{?}{=} -10$$
$$2(-5) = -10 \checkmark$$

Example 1.
$$2(x + 9) = -10$$
$$2x + 18 = -10 \quad \Big| -18$$
$$2x = -28 \quad \Big| \div 2$$
$$x = -14$$

1. Solve the equations. (Give your answer as a fraction, not mixed number, when applicable.)

a. $\quad 5(x + 2) \;=\; 65$	**b.** $\quad 3(y - 2) \;=\; 72$
c. $\quad -8(w - 11) \;=\; 18$	**d.** $\quad 8 \;=\; -9(z + 5)$
e. $\quad 2.5 \;=\; -(6t - 1.5)$	**f.** $\quad -10(-x + 7) \;=\; 27$

Often, you can **start the solution** to an equation in **several different ways**. See the examples below.

Example 2. Solve $4(x + 7) = -36$.

We could start solving this equation by first **applying the distributive property**:

$$\begin{array}{rl|l}
4(x + 7) & = -36 \\
4x + 28 & = -36 & -28 \\
4x & = -64 & \div 4 \\
x & = -16
\end{array}$$

Or, we could **start by dividing** both sides by 4. This ends up being the shorter way, in this case. However, both ways, you will arrive to the correct answer.

$$\begin{array}{rl|l}
4(x + 7) & = -36 & \div 4 \\
x + 7 & = -9 & -7 \\
x & = -16
\end{array}$$

Example 3. Solve $7x - 10 = 45$.

We could solve this equation by first adding 10, then dividing by 7:

$$\begin{array}{rl|l}
7x - 10 & = 45 & +10 \\
7x & = 55 & \div 7 \\
x & = 55/7
\end{array}$$

What if you divide first?

Yes, you can do that. The solution below shows the steps if you divide by 7 first. Notice that 10, too, on the left side <u>has to be divided</u> by 7.

$$\begin{array}{rl|l}
7x - 10 & = 45 & \div 7 \\
x - 10/7 & = 45/7 & +10/7 \\
x & = 45/7 + 10/7 \\
x & = 55/7
\end{array}$$

Comparing the two solutions, we notice that in this case it is easier to subtract first, then divide, because that way we avoid dealing with fractions until the last step.

2. Solve in two ways: (i) by dividing first and (ii) by distributing the multiplication over the brackets first.

Divide first:	Distribute the multiplication first:
$3(x - 7) = 42$	$3(x - 7) = 42$

3. Find two ways to start the solution, and solve each equation in those two ways.

a. Way 1: $-20(q - 5) = 80$	**Way 2:** $-20(q - 5) = 80$
b. Way 1: $-5x - 8 = 27$	**Way 2:** $-5x - 8 = 27$

4. Fill in the margin what is to be done to both sides of the equation, as the next step.

a. $\quad -8(x + 40) = 10$ $\qquad x + 40 = -1.25$ $\qquad\qquad x = -41.25$	**b.** $\quad \dfrac{4x + 2}{7} = -6$ $\qquad 4x + 2 = -42$ $\qquad\quad 4x = -44$ $\qquad\quad x = -11$
c. $\quad 2.3 = -2x + 0.9$ $\qquad 1.4 = -2x$ $\quad -2x = 1.4$ $\qquad x = -0.7$	**d.** $\quad 90 = 20(x - 3)$ $\qquad 90 = 20x - 60$ $\quad 150 = 20x$ $\quad 20x = 150$ $\qquad x = 15/2$

138

5. Use these equations for more practice, as necessary.

a. $\quad -2(x+4) = 7$	**b.** $\quad 9 = 0.3(x-5)$	**c.** $\quad 500 = -20y - 50$
d. $\quad -19 - 4w = 3$	**e.** $\quad -0.5 = 0.2x - 1.4$	**f.** $\quad 5(-x-8) = -10$
g. $\quad 40(3-q) = 65$	**h.** $\quad 200 = -7.5x - 40$	**i.** $\quad 9 - 2s = 78$

a. For what value of a will the equation $x + a = 9$ have -5 as a root?

Puzzle Corner

b. For what value of b will the equation $3x + b = 9$ have -5 as a root?

Equations Review, Part 3

If an equation involves fractions, it is often easier to solve it if you first get rid of them. We do that by **multiplying** both sides of the equation **by the denominator of the fraction** (or by the LCM of the denominators). This is not absolutely necessary as a starting point, but it does make things much easier.

Example 1.

$$\frac{3}{4}a + 4 = 6 \qquad \Big| \cdot 4$$

Note: the *entire* left side needs to be multiplied by 4. That is why we enclose it in brackets.

$$4(\frac{3}{4}a + 4) = 4 \cdot 6$$

$$3a + 16 = 24 \qquad \Big| -16$$

$$3a = 8 \qquad \Big| \div 3$$

$$a = 8/3$$

Check:

$$\frac{3}{4} \cdot \frac{8}{3} + 4 \overset{?}{=} 6$$

$$\frac{8}{4} + 4 \overset{?}{=} 6$$

$$6 = 6 \quad \checkmark$$

Example 2.

$$-\frac{2}{5}(x + 7) = -6 \qquad \Big| \cdot 5$$

$$5 \cdot \left(-\frac{2}{5}\right)(x + 7) = 5(-6) \qquad \text{Next we simplify } 5 \cdot (-2/5).$$

$$-2(x + 7) = -30 \qquad \Big| \div (-2)$$

$$x + 7 = 15 \qquad \Big| -7$$

$$x = 8$$

Check:

$$-\frac{2}{5}(8 + 7) \overset{?}{=} -6$$

$$-\frac{2}{5}(15) \overset{?}{=} -6$$

$$-6 = -6 \quad \checkmark$$

1. Find the errors in these solutions, and correct them.

a.

$$\frac{3}{8}y - 7 = 2 \qquad \Big| \cdot 8$$

$$3y - 7 = 16 \qquad \Big| +7$$

$$3y = 23 \qquad \Big| \div 3$$

$$y = 23/3$$

b.

$$4(y + 2) = \frac{13}{5} \qquad \Big| \cdot 5$$

$$4y + 8 = 13 \qquad \Big| -8$$

$$4y = 5 \qquad \Big| \div 4$$

$$y = 5/4$$

2. Solve the equations. Compare the three and how they are solved.

a. $\frac{1}{5}a + 7 = 3$	b. $\frac{1}{5}(a + 7) = 3$	c. $-\frac{2}{5}(a + 7) = 3$

3. Practice some more. Solve the equations.

a. $2 = -\frac{9}{10}(4 - x)$	b. $2(1 - x) = \frac{5}{12}$	c. $2y - 5 = -\frac{4}{7}$

4. Solve equations involving decimals, also. Use a calculator. Give your final answer rounded to two decimals.

a. $0.4(x + 5) = -3.7$	b. $4.72w - 8.9 = 20$	c. $98.5 = -3(y + 25.6)$

Example 3. Here, the fraction is in a different spot in the equation. Multiplying by the denominator still works.

However, you could also start the solution process by applying the distributive property on the left side.

$$2(x + \tfrac{4}{5}) = -7 \qquad \Big| \cdot 5$$

$$5 \cdot 2(x + \tfrac{4}{5}) = -35$$

$$10(x + \tfrac{4}{5}) = -35$$

$$10x + 8 = -35 \qquad \Big| -8$$

$$10x = -43 \qquad \Big| \div 3$$

$$x = -\tfrac{43}{10}$$

5. Solve the equation from example 4 again, this time starting the solution by applying the distributive property on the left side.

   Hint: don't convert improper fractions to mixed numbers during the solution process. It is easier to calculate with fractions than with mixed numbers.

$$2(x + \tfrac{4}{5}) = -7$$

6. Solve. Compare the three and how they are solved. Again, keep any improper fractions during the process.

a. $-3(x + \tfrac{1}{6}) = 1$	**b.** $-3x + \tfrac{1}{6} = 1$	**c.** $-3x + 1 = -\tfrac{1}{6}$

7. Fill in the missing parts — either what is to be done in the next step, or the missing numbers.

$$2y - 7 = \tfrac{5}{9} \qquad \Big| \; \boxed{}$$

$$9 \cdot (2y - 7) = \boxed{}$$

$$18y - \boxed{} = \boxed{} \qquad \Big| \; \boxed{}$$

$$18y = 68 \qquad \Big| \; \boxed{}$$

$$y = \boxed{}$$

8. **a.** Verify that $x = -4/3$ is *not* a root of this equation.

$$6\left(x - \tfrac{2}{3}\right) = -2$$

$$6x - \tfrac{12}{3} = -12$$

$$6x - 4 = -12$$

$$6x = -8$$

$$x = -8/6 = -4/3$$

b. Find the mistake in the solution, and correct it.

9. Here's a riddle to discover by solving the equations. Use blank paper if needed.

T $3\left(x + \tfrac{2}{9}\right) = -3$	**R** $2 = \tfrac{1}{8}(7 - x)$	**A** $-3x + 6 = \tfrac{3}{5}$
H $0.2(6 - s) = 50$	**E** $1.5 = 3(-T + 0.7)$	**W** $40 - 0.9x = 35.5$

Everyone always talks about it, but no one does anything about it. What is it?

The

5	0.2	9/5	−11/9	−244	0.2	−9

143

Combining Like Terms

1. Try this! Can you simplify the following expressions before you read the lesson material? Don't worry if you can't — this is what we will be learning in this lesson.

a. $4x - 5 - 8x + 11x$	**b.** $2a \cdot 11a$
c. $-2s + 8 - (-3s)$	**d.** $-\dfrac{1}{2}x - \dfrac{3}{4}x + 5$

Remember? A **term** is an expression that consists of numbers and/or variables that are *multiplied*. For example, $-7x$ and $0.6mn^2$ are terms. A single number or variable is also a term. If the term is a single number, such as 4.5 or ¾, we call it a **constant**.

If two terms have the same variable part, they are called **like terms**. For example, $6a$ and $8a$ are like terms, and so are x^2 and $9x^2$, because their variable parts are identical.

But $5x$ and 6 are *not* like terms. The first one has a variable part of x, and the other has no variable part.

We can combine like terms: we can add and subtract them. To do that, simply add/subtract their coefficients. The variable part does not change. Study the examples.

Example 1. $z^3 + 2z^3 - 5z^3$ simplifies to $-2z^3$. (Calculate $1 + 2 - 5 = -2$.)	**Example 2.** $r - t$ cannot be simplified, because the two terms are not like terms.

Example 3. Simplify $6y - 8 - 9y + 2 - 7y$.

First, we organize the expression so that the terms with y are written first, followed by the constant terms.

For that purpose, we **view each operation symbol (+ or −) in front of the term as the *sign* of each term.** In a sense, you can imagine each plus or minus symbol as being "glued" to the term that follows it. Of course the first term, $6y$, gets a "+" sign.

$$+6y - 8 - 9y + 2 - 7y$$

> **Why are we allowed do it this way?**
>
> First, note that subtracting a term is the same as adding its opposite. In symbols,
>
> $$6y \quad -8 \quad -9y + 2 \quad -7y$$
> $$= 6y + (-8) + (-9y) + 2 + (-7y).$$
>
> In other words, the original expression is the SUM of the terms $6y$, -8, $-9y$, 2, and $-7y$. Since addition can be done in any order, we are free to reorganize the terms.

After reordering the terms, the expression becomes $6y - 9y - 7y - 8 + 2$.

Now we need to combine the like terms $6y$, $-9y$, and $-7y$. We do that by finding the sum of their coefficients 6, −9, and −7. Since $6 - 9 - 7 = -10$, then $6y - 9y - 7y = -10y$.

Similarly, we combine the two constant terms: $-8 + 2 = -6$.

Our expression therefore simplifies to $-10y - 6$.

Example 4. $5x \cdot 2x$ simplifies to $10x^2$. Actually, $5x \cdot 2x$ is already a single term, since it only contains multiplication as an operation. So, while we can simplify it by performing the multiplication, this is not a case of combining like terms.

2. Simplify the expressions.

a. $-8m - 6 + 5m - 9$	**b.** $x - 2x + 9x$
c. $-20q - 5q - 19n + 3n$	**d.** $8y - 9 + 7x + 6 - 15y - 2x$
e. $11 - 9m^2 + 2m^2 + 9 - 3.5m^2$	**f.** $16a + 15c + 10d - 7$
g. $2x \cdot x \cdot y \cdot 6y$	**h.** $2x \cdot x \cdot x - 8x - 5x \cdot x^2$
i. $9 - \frac{1}{7}w - \frac{4}{5}w$	**j.** $-\frac{1}{3}x - 2 + \frac{1}{8}x + 5$
k. $\frac{3}{5}n^2 - \frac{2}{3}n^2 - 3m$	**l.** $2x - \frac{1}{2}x^2 - \frac{7}{8}x + 3x^2$

Just like with integers, when we add or subtract variables or expressions, we can use these shortcuts.

Shortcuts for simplification (addition/subtraction only):

$--$ equals a single plus sign: $n - (-m) = n + m$

$++$ equals a single plus sign: $n + (+m) = n + m$

$+-$ equals a single minus sign: $n + (-m) = n - m$

$-+$ equals a single minus sign: $n - (+m) = n - m$

3. Simplify the expressions.

a. $2 + (-s)$	**b.** $5 - (-x)$	**c.** $x + (+4) + (-3x)$
d. $2x^2 - (-y^2)$	**e.** $6.2w + (-1.3w)$	**f.** $7w + (-2w) - (+4w)$
g. $7x - (-2x) + 1 - 9x + 3 + (-4x)$		**h.** $2 - 4.2y + (-y) + 0.5y + 1.6 - (-1.5y)$
i. $-60x + 80 - (-x) - 30x + 30$		**j.** $5 - (-2z) - (-z) + (-2z) - 8 + 20z$

When solving equations, you often need to combine like terms to simplify one side of the equation.

Example 5. Solve $2x - 3x + 9x = -24$.

There are several terms with x on the left side, so we will first combine them. Then the equation is transformed into an easy, one-step equation.

$$2x - 3x + 9x = -24 \quad \text{(combine like terms)}$$
$$8x = -24 \quad \text{(divide both sides by eight)}$$
$$x = -3$$

To check the equation, substitute the root -3 in place of the unknown in the *original* equation:

$$2 \cdot (-3) - 3 \cdot (-3) + 9 \cdot (-3) \overset{?}{=} -24$$
$$-6 - (-9) + (-27) \overset{?}{=} -24$$
$$-6 + 9 - 27) \overset{?}{=} -24$$
$$-24 = -24 \quad \checkmark \quad \text{It checks!}$$

4. Solve each equation. Start out by combining like terms on one side of the equation.

a. $\quad 5x + 2x = 14$	**b.** $\quad -7y + 3y - (-2y) = 50$
c. $\quad -2y + (-5y) + 8y - 7 = 8$	**d.** $\quad 20 - 36 = -8x + 9x - 6x$
e. $\quad 1.5s - (-4.8s) - 1.3s = 3.5$	**f.** $\quad 3.2 = -(-2x) - x - 5x + 1.6$
g. $\quad 2t - 4.8t + 1.3t - 0.8t = 4.6$	**h.** $\quad 11 = \frac{1}{2}x - \frac{1}{8}x + 5$

Word Problems

Example 1. The width of a rectangle is three times its length, and its perimeter is 28 m. What are the width and the length of the rectangle?

First, check *what quantity* is unknown, and choose a variable for it.

We don't know the length nor the width of this rectangle. However, since we know that the width is three times the length, we can write the width as an expression of the length. In other words, we can let x be the length, and then the width is $3x$. We don't need another variable for the width.

In your solution, **you need to specify what x denotes.** You can write: "Let x be the length of the rectangle." After that, write something to the effect: "Then, $3x$ is the width." This is to ensure that a reader understands your thought process.

Continuing on with the solution, we know the perimeter is the sum of all four sides, and equals 28 m. From that we can write the equation: $x + 3x + x + 3x = 28$. Solving that will be easy!

1. Solve the equation in example 1. Lastly, answer the actual question
   (specify what the width and the height of the rectangle actually are).

2. The length of a certain rectangle is 12 cm more than its width. The perimeter is 136 cm. How much do the sides of the rectangle measure? Write an equation.

   <u>Reminder:</u> Note clearly what your chosen variable denotes in the problem.

Example 2. Susan and Henry divide a task of washing 170 windows in a large building in a ratio of 3:2. How many windows will Henry wash?

You have learned to solve this type of problem using a bar model. We can also use an equation (which will exactly match the basic idea in the bar model).

Let $3x$ be the amount of windows that Susan washes, and $2x$ the amount that Henry washes. This way, the windows they washed are in the ratio of 3:2. Since we know they washed a total of 170 windows, we can write the equation $3x + 2x = 170$.

3. Solve the equation in example 2. How many windows will Henry wash?

4. An inheritance of $354 000 was divided between three heirs in the ratio of 1:6:5. How much did each heir get?

5. The two sides of a rectangle are in a ratio of 3:5 and its perimeter is 416 cm. What are the dimensions of the rectangle?

> **Example 3.** Another type of problem you have solved previously using a bar model is where the total is known, and the parts making up the total differ by a known amount.
>
> Eric and Jeremy worked last week for a total of 99 hours, and Eric worked 5 more hours than Jeremy. How many hours did Eric work?
>
> Let x be Jeremy's working hours. Then, Eric worked for $x + 5$ hours. Knowing the total, we can easily write an equation for this situation.

6. Write an equation for Example 3, and solve it. Lastly, answer the actual question that was asked.

7. Anna has a flock of chickens and a flock of ducks. She has 17 more chickens than ducks, and in total she has 135 birds. How many chickens does she have?

8. Hans has to cross a bridge and pay a $6 bridge toll when going to work (coming back, he doesn't have to pay it). Some days he carpools with two other people, and they share the toll fee equally.

   a. Hans noticed that in the last two weeks (which had 10 workdays), he had paid a total of $36 in bridge tolls. How many days of those 10 did he carpool?

   b. In a month (which had 22 work days), he had paid a total of $96 in bridge tolls. How many days of those did he *not* carpool?

9. The pet store had three different leashes for sale. The price of one was $5.40 more than the price of the cheapest, and the price of another was $11.60 more than the cheapest. If you bought all three, the total came to $62.60. How much did the most expensive leash cost?

10. The sum of three consecutive whole numbers is 360. What are the numbers?

Hint: Let x be the first of the three consecutive whole numbers. What are the other two, in terms of x?

11. The sum of three consecutive odd numbers is 1971. What are the numbers?

Hint: Let x be the first of the three consecutive odd numbers. What is the next odd number, in terms of x?

12. The sum of four consecutive multiples of 5 is 1570. What are the numbers?

It is possible to write a set of equations for the following problems, but you haven't studied those types of equations (quadratic) yet. So, use **guess and check**. It can work equally efficiently!

a. The sum of two numbers is 35, and their product is 300. What are the numbers?

b. The sum of two numbers is 220, and their product is 9600. What are the numbers?

Puzzle Corner

A Variable on Both Sides

Example 1. Solve $2x + 8 = -5x$.

Notice that the unknown appears on both sides of the equation. This is not a problem; we can still use the principle of doing the same operation to both sides in order to isolate the unknown on one side. In this case, we can either subtract $2x$ from both sides or add $5x$ to both sides. See both options below.

First subtract 2x:		First add 5x:		Check:
$2x + 8 = -5x$ $\quad$ $\vert$ $- 2x$		$2x + 8 = -5x$ $\quad$ $\vert$ $+ 5x$		$2 \cdot (-8/7) + 8 \overset{?}{=} -5 \cdot (-8/7)$
$8 = -7x$ $\quad$ (Switch sides.)		$7x + 8 = 0$ $\quad$ $\vert$ $- 8$		$-16/7 + 8 \overset{?}{=} 40/7$
$-7x = 8$ $\quad$ $\vert$ $\div -7$		$7x = -8$ $\quad$ $\vert$ $\div 7$		$-16/7 + 56/7 \overset{?}{=} 40/7$
$x = -8/7$		$x = -8/7$		$40/7 \overset{?}{=} 40/7$ ✓

1. Solve the equation in two ways, as instructed.

First add 2s:	**First subtract 4s:**
$10 - 2s = 4s + 9$ $\qquad$ $\vert + 2s$	$10 - 2s = 4s + 9$ $\qquad$ $\vert - 4s$

2. Solve. Check your solutions (as always!).

a. $\quad$ $3x + 2 = 2x - 7$	**b.** $\quad$ $9y - 2 = 7y + 5$

3. A common student error is to add or subtract "across the sides," instead of carefully adding or subtracting the same quantity to/from both sides.

Here is an example of it: the student added $7w$ and $2w$, and wrote $9w$ on the next line. Correct the error and solve the equation.

$$7w + 8 = 2w - 5$$

$$9w + 8 = -5$$

4. Solve. Check your solutions (as always!).

a. $-2y - 6 = 20 + 6y$	b. $8x - 12 = -1 - 3x$	c. $6z - 5 = 9 - 2z$

5. Fred is contemplating two different job offers. In one, he gets paid $19.50 per hour plus he will receive a bonus based on the sales he brings in, which he estimates to be about $150 per week. In another job, he will earn $21 per hour (no bonuses).

   a. Write an expression for the weekly earnings in each job, for m hours of work.

   Job 1:   Job 2:

   b. In which job would he earn more, if he worked 20 hours per week?

   c. For what amount of work hours would both jobs provide him the same wages?

Sometimes there are several variable terms on one side. Often, the most effective strategy is to combine the like terms first.

Example 2.

$$20 - 8m + 2m = 6m - 4m - 4 \qquad \text{(Combine like terms, separately on each side.)}$$

$$
\begin{array}{rl|l}
20 - 6m &= 2m - 4 & -2m \\
20 - 8m &= -4 & -20 \\
-8m &= -24 & \div(-8) \\
m &= 3 &
\end{array}
$$

Check:

$$20 - 8(3) + 2(3) \overset{?}{=} 6(3) - 4(3) - 4$$

$$20 - 24 + 6 \overset{?}{=} 18 - 12 - 4$$

$$2 = 2 \checkmark$$

6. Fill in the missing numbers in the solution. As your first step, combine like terms, separately on each side of the equation. Then finish the solution of the equation.

a. $3x + 7 - 5x = 6x + 1 - 4x$

$$\boxed{}x + 7 = \boxed{}x + 1 \quad \Big| -7$$

$$\boxed{}x = \boxed{}x - \boxed{}$$

$$=$$

b. $-7x - 5 + x + 4x = 8 - 2x - 5x$

$$\boxed{}x - 5 = 8 - \boxed{}x$$

$$=$$

7. Solve. Check your solutions.

a.	b.	c.
$6x + 3x + 1 = 9x - 2x - 7$	$16y - 4y - 3 = -4y - y$	$-26x + 12x = -18x + 8x - 6$

8. Solve. Give the solutions rounded to two decimals.

a.	b.
$0.9y + 1 - 1.4y = 4.6y - 4.8 + y$	$4(0.7w + 0.9) = 1.6 - 0.8w$

9. Use these for extra practice, as necessary. If necessary, round the solution to two decimals.

a. $6w - 6.5 = 2w - 1$	b. $11 - 2q = 7 - 5q$	c. $5g - 5 = -20 - 2g$
d. $2.56x + 2 = 5.1 - 4.89x$	e. $2(14.85z + 0.8) + 2z = 0.5(z - 3)$	

Word Problems and More Practice

1. The area of this two-part rectangle is 253 square feet.
   Its one side is 11 ft and the other is $(9 + x)$ feet.

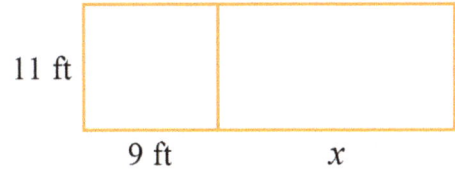

   a. Write an equation for its area.
      Then solve for x in your equation.

   11 ft ▭▭

   9 ft x

   b. Draw a two-part rectangle to represent
      $16(x + 11) = 496$, and solve for x.

2. The perimeter of a certain triangle is five times its shortest side. The second side is 5 units longer than the first, and the third side is 15 units longer than the first. Find the side lengths of the triangle.

3. Study Robert's solutions below. What same type of mistake does he make every time?
   Also, correct the mistakes.

a. $3y + 5 = 8y + 7$	**b.** $5x - 7 = 20 - 3x$
$11y = 12$	$2x = 13$
$y = 12/11$	$x = 13/2$

4. A teacher ordered 24 writing workbooks for her class. The price had been lowered by $3.50 from the original. Her total came to $232.80. What was the price of one workbook before the discount? Choose a variable for the unknown and write an equation for the situation. Solve the equation.

5. Henry bought 40 bags of garden soil, for $13.90 a bag, but he got a discount on half of them. If his total came to $498, find how much the discount was.

   Hint: Write an expression for the total cost.

6. Solve. Check your solutions (as always!).

a. $-50w - 30 = 26w + 18$	b. $12x + 9 = -3 - 5x - 8$	c. $2m - 1 = 9 - 2m - 7 - 8m$

7. Solve. Check your solutions.

a. $8 - 2m + 5 - 8m = 20 - m + 5m - 2m$	**b.** $49.5 - 1.2s = 2.4s - 22.5$
c. $7(50x - 5) = 40x + 145 + 130x$	**d.** $6(2 + 8y) - 12y + 4 = 2(9 - 5y) - 2$

a. The root of the equation below is 3. Find the missing number (the coefficient of x).

$$\boxed{}\, x - 8 = -2$$

Puzzle Corner

b. Find the value of b for which the equation $bx + 5 = -4$ will have a root of $x = 1$.

Simplifying Linear Expressions

Example 1. Simplify $10x + 7(3x - 6)$.

Before we can combine like terms, we need to use the distributive property to remove the brackets. (Why? Because of *the order of operations*!) →

$$10x + 7(3x - 6)$$
$$= 10x + 21x - 42$$
$$= 31x - 42$$

Example 2. Simplify $x - 2(5x + 3)$.

Again, we need to work on the brackets first, and we use the distributive property to do so. Notice carefully how this is done, and especially pay attention to the minus sign in front of the 6. →

Where does this minus come from?

It comes from taking the -2 through the brackets: $-2 \cdot 3 = -6$.

$$x - 2(5x + 3)$$
$$= x - 10x - 6$$
$$= -9x - 6$$

Example 3. Simplify $7x + 9 - (8x - 3)$.

This time, the highlighted subtraction acts the same as a multiplication by -1, and both terms inside brackets ($8x$ and -3) become their *opposites.* →

$$7x + 9 - (8x - 3)$$
$$= 7x + 9 - 8x + 3$$
$$= -x + 12$$

$$7x + 9 - (8x - 3)$$
$$= 7x + 9 + [-(8x - 3)]$$
$$= 7x + 9 + [-8x + 3]$$
$$= 7x + 9 + (-8x) + 3$$
$$= -x + 12$$

Remember that if the process on the right seems convoluted, you can go a longer route, as shown on the left. You can change the subtraction to an addition of the opposite.

(We use square brackets instead of nested brackets for clarity.)

1. These simplifications show a common student error.

    a. What is it?

    b. Correct the errors.

$$9x - 3(x - 4)$$
$$\downarrow$$
$$9x - 3x - 12$$

and

$$20 - (6 - 9y)$$
$$\downarrow$$
$$20 - 6 - 9y$$

2. Simplify.

a. $4x + 9(2x - 3)$	**b.** $4x - 9(2x - 3)$	**c.** $5x - (8x + 3)$	**d.** $5x - (8x - 3)$
e. $4y - 80 + 9(20 - 8y)$	**f.** $w - 7 - (-3 + 6w)$	**g.** $-5s - (8s - 3)$	**h.** $1 + x - 3(7x - 3)$

Example 4. Let A be the expression $-2x + 9$ and B be the expression $5x - 3$. Write the sum of these two expressions, and simplify it.

First let's write down A + B. It is simply $(-2x + 9) + (5x - 3)$.

To simplify it, we first remove the brackets and then combine like terms. →

The first set of brackets can be removed without any changes. The second can also, since there is only a + sign in front of it. (If it was minus, it would be a different story.) See the full process of simplification on the right.

$$(-2x + 9) + (5x - 3)$$
$$= -2x + 9 + 5x - 3$$
$$= 3x + 6$$

3. Let $A = 12x - 9$ and $B = 3x - 8$.

   a. Write the sum of A and B and simplify it.

   b. Write the difference of A and B and simplify it.

   c. Write the sum 2A + B and simplify it.

4. Simplify.

a. $5(9 - 2y) + 7(3 - 5y)$	**b.** $5(6x - 12) + 7(x + 9)$
c. $5(1.2 - 0.8x) - 10(0.5x - 4.2)$	**d.** $2(0.9w - v) - 4(w + 0.5v)$
e. $\frac{1}{4}(60v - 16) - \frac{2}{3}(45v + 15)$	**f.** $-\frac{1}{2}(24x + 6) - 5(3x + \frac{1}{5})$
g. $36s - 9(3s + \frac{2}{3})$	**h.** $-\frac{1}{2}(30y + 12 - 6y) - 15y + 9$

Example 5. The skills you have practiced in this lesson come in handy when solving equations.

Check:

$$2[3(2) + 1] - 7(2) \overset{?}{=} 30 - 5(2 + 4)$$

$$2[6 + 1] - 14 \overset{?}{=} 30 - 5(6)$$

$$2(7) - 14 = 0 \checkmark$$

$2(3x + 1) - 7x = 30 - 5(x + 4)$	(Apply the distributive property.)
$6x + 2 - 7x = 30 - 5x - 20$	(Combine like terms & simplify.)
$2 - x = 10 - 5x$	$+\,5x$
$2 + 4x = 10$	$-\,2$
$4x = 8$	$\div\,4$
$x = 2$	

5. Solve.

a. $4(x - 2) = 8(x + 1) + 20x$	**b.** $10y + 2(y - 6) = 3y + 5(4 - 2y)$
c. $-2(y - 12) = 30 - 6(y + 4)$	**d.** $30(-5x - 7) + 20x = 80x - 5(10x + 1)$

More Practice

1. Theresa made an error in simplification.

   a. On which line is her error?

   b. Correct her error.

(1)	$-(14a + 11) - \dfrac{3}{4}(12a - 32)$
(2)	$-14a - 11 - \dfrac{3}{4}(12a - 32)$
(3)	$-14a - 11 - 9a - 24$
(4)	$-23a - 35$

2. **a.** Write the expression below as a product of two factors:

   $200x - 80y + 420$

   b. Is there another solution?

3. Solve.

   a. $\quad 1.5(x - 2) \;=\; 0.6(x + 4) + x$

   b. $\quad 10y - 2(y - 6) \;=\; 3y + 5(4 - 2y)$

   c. $\quad 1060 - 20(a - 10) \;=\; 50a$

   d. $\quad -\dfrac{1}{2}(36x + 12) \;=\; \dfrac{1}{3}(3x + 24 - 12x)$

4. Jerry bought two different kinds of geese, 15 geese in total. Some cost $29, and the rest were twice that price. His total came to $725. How many cheaper geese did he buy?

   Hint: write an expression for the total cost.

5. Fill in the pyramid!
   Add each pair of expressions in neighbouring blocks, and write their sum in the block directly above it.

$$x + 3 \qquad 5x - 4 \qquad 7 - 3x \qquad -6x$$

6. Write an expression for the sequence of operations. Then simplify each expression.

   a. Add 4 to $5x$, then double what you have, and lastly subtract 1.

   b. Add 4 to $5x$, subtract 1 from the result, and lastly double the entire result.

   c. The difference between the quantities $4x - 5$ and $10x + 9$.

   d. Take the sum of 0.3 and $4x$ two times, then subtract that from 1.

   e. Subtract the sum of 0.3 and $4x$ from 1, and lastly double what you have.

7. Solve. Check your solutions. In (b), give your answer rounded to two decimal digits.

a.	b.
$-12 - 4(z - 3) = 5z + 8$	$4.5w + 0.8(-2w + 4) = 4 - 0.5(3 - w)$

8. In the town of Dreamy, seven different months last year had the same number of rainy days.
   The rest of the months were wetter, and had five more rainy days each than the less rainy months.
   In total, the town had 109 rainy days that year.

   How many rainy days were there in each of the less rainy months?

9. A wise father's will read: "I bequeath my possessions to my three sons as they have worked for me when growing up. My middle son will get half as much as my oldest son, and my youngest will get three times as much as my oldest."

   If the inheritance was worth $1 215 000, calculate how much was each son's share of it.

Age and Coin Word Problems

> **Example 1.** Ann says, "In 10 years, I will be twice as old as I was 22 years ago. How old am I now?"
>
> To solve this little puzzle, we need to first choose a quantity as an unknown. What is not known? Obviously, Ann's age now. So, let x be Ann's age now. (Remember always to explain what quantity or thing your unknown denotes.)
>
> Now we translate the clue given to us into symbols. First, her age 22 years ago is $x - 22$. Her age in 10 years is $x + 10$. Twice as old as 22 years ago is $2(x - 22)$. The clue translates into the equation
>
> $$x + 10 = 2(x - 22)$$

1. Solve the equation from example 1.

2. Harry says, "Five years ago, I was 1/3 as old as I will be in 45 years." How old is Harry?

3. Hannah says, "In 18 years, I will be three times as old as I was ten years ago."
   How old is Hannah?

4. Fifteen years ago, my age was only half of what it will be in 3 years. How old am I?

Example 2. A cashier has a total of 153 bank notes, in tens, twenties, and fifties. She has twice as many tens as twenties, and 15 less fifties than twenties. How many of each denomination of notes does she have?

Since the quantity of tens and fifties are given in terms of twenties, it's easiest to choose the number of 20-dollar notes as our unknown. So, let y be the number of 20-dollar notes. Then:

- The number of 10-dollar notes is $2y$.
- The number of 50-dollar notes is $y - 15$.

Your turn to think: how do you write an actual *equation* to fit the situation?

5. Write an equation for the situation in Example 2, and solve it.

6. Mom has a lot of coins saved for helping her children learn how to count coins. She has half as many 50-cent pieces as dimes, twice as many nickels as dimes, and four less quarters than dimes. The total number of coins she has is 59. How many of each type of coin does she have?

7. Joe, Jill, and Jack worked together on a job. Jill earned $150 more than Joe, and Jack earned $70 less than Joe. Together they earned $1280. How much did each person earn?

Example 3. Ryan has a bunch of quarters and dimes. He has 15 more dimes than quarters, and the total *value* of his coins is $5.35. How many quarters does he have?

Let x be the number of quarters that he has. Then, the number of dimes is $x + 15$. To take into account the *value* of the coins, we multiply those by 0.25 and 0.1:

- The value of his quarters is $0.25x$.
- The value of the dimes is $0.1(x + 15)$.

Now we can write an equation relating the total value of his coins — a task for you!

8. Write an equation for the situation in Example 3, and solve it. Also, answer the question.
   Hint: If you'd like, you can get rid of the decimals in your equation by multiplying both sides by something.

9. Sheila's piggy bank has a bunch of quarters, no nickels, a third as many dimes as quarters, and four 50-cent pieces. The total value of the contents is $12.20. How many of each type of coin does she have?

10. Greg gives his brother a little puzzle about the bank notes he has, "I have three times as many tens as twenties, half as many fifties as twenties, and two hundreds. The total value of my money is $1550." How many of each type of bank note does he have?

Equations with Fractions 1

When solving an equation that has denominators (fractions or fractional expressions), the usual, and often the most efficient way, is to multiply both sides of the equation by the least common multiple (LCM) of the denominators. You can also use *any* multiple of the denominators, if you cannot find the LCM.

Example 1. Here, it makes sense to multiply the equation by the LCM of the denominators (4 and 8), which is **8**. This eliminates both fractions, and leaves us with an equation with only whole numbers, on line 3.

1	$\frac{1}{4}a + 5 = \frac{3}{8}$	$\cdot \, 8$
2	$8(\frac{1}{4}a + 5) = 8 \cdot \frac{3}{8}$	Note the *entire* left side needs to be multiplied by 8, thus the need for brackets.
3	$2a + 40 = 3$	$-\, 40$
4	$2a = -37$	$\div \, 2$
5	$a = -37/2$	

Check: $\frac{1}{4} \cdot (-\frac{37}{2}) + 5 \overset{?}{=} \frac{3}{8}$

$(-\frac{37}{8}) + 5 \overset{?}{=} \frac{3}{8}$

$-4\frac{5}{8} + 5 = \frac{3}{8}$ ✓

1. Guided practice! Finish solving each equation. Lastly, check your solution.

a. Here again, multiplying both sides by 10 (the LCM of the denominators 2 and 5) eventually eliminates both fractions, though it takes more than one step.

$$\frac{3}{5}x + \frac{1}{2} = -3 \qquad \Big| \cdot 10$$

$$10(\frac{3}{5}x + \frac{1}{2}) = 10 \cdot (-3)$$

$$=$$

$$=$$

$$=$$

b. Here we have three fractions. Multiplying both sides by 12 eventually eliminates all fractions, though it takes more than one step.

$$\frac{2}{3}x - \frac{1}{6} = -\frac{1}{4} \qquad \Big| \cdot 12$$

$$12 \cdot (\frac{2}{3}x - \frac{1}{6}) = 12 \cdot (-\frac{1}{4})$$

$$=$$

$$=$$

$$=$$

Whenever you multiply an equation by a number, *each term* needs multiplied by that number. (Why?) A common student error is to miss multiplying some terms.

Example 2. Here, we multiply the equation by the LCM of the denominators (4 and 6), which is 12. Notice that **all four terms** get multiplied by that 12.

1   $\frac{1}{4}x + 5 = \frac{1}{6}x - 7$ $\Big| \cdot 12$

2   $12(\frac{1}{4}x) + 12 \cdot 5 = 12(\frac{1}{6}x) - 12 \cdot 7$ Each term is multiplied by 12 separately. As you gain more experience, you can skip writing this step.

3   $3x + 60 = 2x - 84$

4   $x = -144$

Example 3. On the right, it looks like the terms x and 5 do not get multiplied — yet this is correct. How can that be?

$\frac{1}{4}(x + 5) = 10 - x$ $\Big| \cdot 4$

$x + 5 = 40 - 4x$

It is because on the left side, we have ¼ times a quantity.

When this gets multiplied by four, the four and ¼ *cancel each other*: $4 \cdot \frac{1}{4}(x + 5)$

$\downarrow$

$1(x + 5)$

$\downarrow$

$x + 5$

2. Multiply each expression by 5 and simplify. (These are *not* equations.)

a. $\frac{1}{5}x - 4$	b. $\frac{1}{5}(x - 4)$	c. $\frac{2}{5}(x - 3)$	d. $\frac{2}{5}x - 3$
Multiplied by 5:	Multiplied by 5:	Multiplied by 5:	Multiplied by 5:
$5(\frac{1}{5}x - 4) =$			

3. Find the errors in these solutions, and correct them. Finish the solution in (b).

a. $\frac{3}{4}x - 0.5 = \frac{1}{10}x + 1$ $\Big| \cdot 20$

$15x - 10 = 2x + 1$

$13x = 11$

$x = 11/13$

b. $2(y - 5) = \frac{3}{5}(y - 6)$ $\Big| \cdot 5$

$10(y - 25) = 3y - 30$

4. Solve the equations. *Hint:* use a calculator to check the solution. If the root is a fraction, it is often helpful to use a decimal approximation of it, and check that both sides of the original equation are approximately equal.

a. $4x - \frac{1}{10} = \frac{3}{5}x - 6$	**b.** $\frac{1}{6}x - 4 = \frac{2}{3}x - 1$	**c.** $\frac{1}{3}(x - 4) = -\frac{1}{8}$
d. $-\frac{1}{2} = \frac{1}{10}(x + 5)$	**e.** $2(x - \frac{5}{8}) = x - \frac{1}{2}$	**f.** $\frac{1}{2}(x - \frac{3}{4}) = 10$

5. You might wonder what happens if you don't multiply by the LCM of the denominators first. Let's find out!

This is the equation from example 1. This time, start the solution by subtracting 5 from both sides.

$$\frac{1}{4}a + 5 = \frac{3}{8}$$

$$=$$

Equations with Fractions 2

Example 1. Also in this situation, it makes sense to start by multiplying out the denominators. We use 12 since it is a common multiple of both 3 and 4.

Notice how things simplify in the next step.

Checking the solution, we get:

$$\frac{11.5 + 5}{3} \overset{?}{=} \frac{2(11.5) - 1}{4}$$

$$\frac{16.5}{3} \overset{?}{=} \frac{23 - 1}{4}$$

$$5.5 = 22/4$$

$$\frac{x + 5}{3} = \frac{2x - 1}{4} \quad \bigg| \cdot 12$$

$$12 \cdot \left(\frac{x + 5}{3}\right) = 12 \cdot \left(\frac{2x - 1}{4}\right) \quad \text{(Simplify.)}$$

$$4(x + 5) = 3(2x - 1)$$

$$4x + 20 = 6x - 3 \quad \bigg| - 6x$$

$$-2x + 20 = -3 \quad \bigg| - 20$$

$$-2x = -23 \quad \bigg| \div 2$$

$$x = 23/2$$

1. Solve. Can you think of two different ways to start the solution? *Hint:* Again it will be handy to check the solutions with a calculator using a decimal approximation of the root.

a. $\quad \dfrac{3x - 4}{2} = \dfrac{3x + 1}{5}$	**b.** $\quad \dfrac{15 - 2s}{8} = \dfrac{5s - 1}{2}$

2. What errors are made in these solutions? Correct them, and continue the solutions.

a. $\quad \dfrac{3x - 4}{2} - 5 = 7 \quad \bigg| \cdot 2$

$$3x - 4 - 5 = 14$$

b. $\quad 3 - x = 2x + \dfrac{x - 10}{2} \quad \bigg| \cdot 10$

$$30 - x = 2x + 5x - 50$$

3. Solve. What is different about the two equations (a), and (b)? How does that affect the solution process?

a. $2x + \dfrac{5-x}{6} = 4$

b. $2x - \dfrac{5-x}{6} = 4$

4. Practice some more!

a. $\dfrac{3x-8}{10} - 1 = x$

b. $11 = 3y + \dfrac{5-5y}{3}$

c. $0 = \dfrac{3x-2}{4} + \dfrac{x+2}{5}$

d. $-x + \dfrac{1-3x}{2} = \dfrac{x}{3} + 2$

5. Solve equations involving decimals, also. Hint: In (c), you can cross-multiply. Use a calculator. Give your final answer rounded to two decimals.

| a. $\dfrac{3.2x - 1}{5} = 0.9x$ | b. $0.08x - \dfrac{0.1x}{4} = 0.2$ |
| c. $\dfrac{20x - 4.3}{0.4} = \dfrac{3.89x}{2.5}$ | d. $5.4 - \dfrac{0.3 - x}{4} = \dfrac{x}{2}$ |

6. Check what happens if you start the solution of this equation by multiplying both sides by **5** (not by 10).

$$\frac{3}{5}\left(x + \frac{1}{2}\right) = -3$$

$$=$$

Andrea put forth a puzzle: "The sides of my rectangle are consecutive whole numbers, and the area is between 3200 and 3400 square units." What are the sides of Andrea's rectangle?

Formulas, Part 1

Example 1. To calculate the area of a triangle, we can use the

formula $A = \dfrac{bh}{2}$, where b is the base and h is the height of the triangle.

This formula has three variables: A, b, and h. Treating the formula as an equation, we can solve it for any one of those variables. Above you see it solved for A, or the area. Let's solve it for h.

This means we treat h as the unknown, and the other letters as if they were known numbers (constants). So, we want to isolate h. See the process on the right.

$$A = \frac{bh}{2} \quad \Big| \cdot 2$$
$$2A = bh \quad \Big| \div b$$
$$\frac{2A}{b} = h$$
$$h = \frac{2A}{b}$$

You may use a calculator in all the problems in this lesson.

1. Use the formula for the area of a triangle as solved for h to answer the question.
   If an area of a triangle is 250 cm^2, and its base is 50 cm, what is its height?

2. **a.** The volume of a circular prism is $V = \pi r^2 h$,
   where r is the radius of the circle and h is its height.
   Solve this for h.

   b. What is the height of a circular prism with a
   volume of 8.00 m^3 and a radius of 60.0 cm?

3. The formula $m = \dfrac{a_1 + a_2 + a_3}{3}$ gives you the mean (average)
   of three numbers a_1, a_2, and a_3.

   a. Solve this formula for a_1.

   b. Jennifer has gotten 78 and 82 points on two English tests, and she
   has one more to go. She wants her average for the three tests to be
   85 (at least). What should she get on her third test in order to achieve that?

If you are given the values for all the variables in a formula except one, you have two options:

1. You can do like we did on the previous page: first solve the formula for the unknown, then substitute the known values.

2. Or, first substitute the known values into the formula, then solve for the unknown variable.

See the example below for how option (2) works.

Example 2. The volume of a cone is $V = \dfrac{A_b h}{3}$, where A_b is the area of the base and h is the height of the cone. The volume of a cone is 450 cm^3, and its height is 25 cm. What is the area of its base?

Substituting the known values into the formula, we get
$450 \text{ cm}^3 = \dfrac{A_b \cdot 25 \text{ cm}}{3}$. See the full solution on the right:

$$450 \text{ cm}^3 = \frac{A_b \cdot 25 \text{ cm}}{3} \quad \Big| \cdot 3$$

$$1350 \text{ cm}^3 = 25 \text{ cm} \cdot A_b \quad \Big| \div 25 \text{ cm}$$

$$\frac{1350 \text{ cm}^3}{25 \text{ cm}} = A_b$$

$$A_b = 54 \text{ cm}^2$$

4. The formula $d = vt$ gives us the distance (d) travelled with speed (velocity) v in time t.
   How long will it take for you to travel 245 km with a speed of 65 km/h?

5. Sarah drove half of the distance from her home to town at a speed of 60 km/h and the other half at a speed of 40 km/h. If the distance from her home to town is 16 km, calculate how long it took her.

6. The final cost (C) of purchasing m items at an x percent discount, when the normal price is p is given by the formula $C = (1 - x/100)mp$. If an item costs \$15, and you will get an 8% discount, find how many of that item you can get with \$1200.

Formulas, Part 2

You may use a calculator in all the problems in this lesson.

1. The area of a trapezoid is A = $\dfrac{a+b}{2}\,h$, where a and b
   are the two parallel sides, and h is the height.
   Solve this formula for a.

2. **a.** The formula C = $\dfrac{5}{9}$(F − 32) gives the temperature
   in degrees Celsius when the temperature in degrees
   Fahrenheit (F) is known. Solve this formula for F.

   b. A heat wave in Europe! France is experiencing
   temperature highs of 40°C!
   How much is this in Fahrenheit?

3. The fuel consumption (F) of a vehicle can be calculated by the formula **F = g/d**, where g is the
   amount of gasoline used in litres and d is the distance driven in kilometres. This formula will give
   the fuel consumption in litres per km. If you want the answer in litres per 100 km, multiply the
   final result by 100.

   a. Find the fuel consumption of a car that used 28 litres of gasoline for a distance of 350 km.
   Give your answer both in litres/km and in litres/100 km.

   b. A vehicle's fuel consumption is 8.6 L per 100 km. How far will it go with 39 litres of fuel?
   Use logical thinking.

   c. Now solve the formula F = g/d for d, and use that to calculate the answer to the question in (c).

4. **a.** Write a formula for the cost (C) of driving a distance d, using the three variables distance (d), fuel consumption a car (F) in litres per km, and the price of gasoline per litre (p). Use logical thinking!

   b. Use your formula to find the cost of driving a car that uses 7.9 L per 100 km a distance of 600 km, the price of gasoline being $1.80 per litre.

   c. Use your formula to find the distance you can drive a car that uses 7.2 L per 100 km with $100 worth of gasoline, the price of gasoline being $1.92 per litre.

5. Edward invested $5000 to a bank account with a 6% interest rate, for 3 years. How much was in his account at the end of five years?

> The formula for simple interest (I) is $I = Prt$, where P is the principal, r is the interest rate, and t is time.

6. If you want your $2500 to earn $500 in interest over two years, how much should the interest rate be?

7. How long will it take for $12 000 to earn $500 in interest if the interest rate is 8.4%?

Puzzle Corner

Ann wants to invest $10 000, splitting that amount between two different opportunities. One has the annual interest rate of 8% (but has some other disadvantages), and the other has the rate of 6%. How should she split her $10 000 so that she will earn $6000 in eight years?

Hint: It may look like this involves *two* unknowns: the two principal amounts. However, since the two add up to $10 000, in reality we can get by with one.

More on Equations

Notice what happens in the end of the solution process of this equation.

We end up with the equation $0 = 0$. It does not have the variable, but it is a true equation.

Just before that, on line 4, we have the equation $6x = 6x$. *Any* value of the variable x fulfills that!

And if you go backwards to line 3, 2, and 1, any value of x fulfills those equations, too.

Example 1.

1	$4(x - 3) + 2x$	$= 18 - 6x - 30$	
2	$4x - 12 + 2x$	$= 6x - 12$	
3	$6x - 12$	$= 6x - 12$	$+ 12$
4	$6x$	$= 6x$	$- 6x$
5	0	$= 0$	

An equation that is true for all values of the variable(s) in it is called an **identity**. It has an **infinite number of solutions**.

If the solution process leads you to an equation of the form $a = a$, then the original equation is an identity (and all the ones in your solution process are also).

Here, we end up with the false equation $6 = 1$.

This means the original equation has *no solutions*.

You can probably see that even in line 2.

Example 2.

1	$3(x + 2)$	$= 9x + 1 - 6x$	
2	$3x + 6$	$= 3x + 1$	$- 3x$
3	6	$= 1$	

1. Solve the equations. Indicate whether each equation has one, none, or an infinite number of solutions.

a. $2x + 2 = 2x - 7$	**b.** $2x + 1 = 2(x - 3) + 7$	**c.** $5x + 1 = 2(x - 3)$

2. Jorge claims the equation $9x + 5 = 5$ has an infinite number of solutions,
   Is he correct? Why or why not?

An equation with one variable is a **linear equation** if it can be written in the form $ax + b = 0$, where a and b are real numbers and x is the variable. (When plotted, the graph of a linear equation is a line; hence the name.)

All the equations we have dealt with in this book are linear equations.

Every linear equation falls into one of these categories:

- It has **one** unique solution.
- It has **no** solutions.
- It has an **infinite number** of solutions.

3. Solve the equations. Indicate whether each equation has one, none, or an infinite number of solutions.

a. $\quad 10 - 7x \;=\; x - 10$	**b.** $\quad 4x + x \;=\; 5x - 3 + 7$	**c.** $\quad 2 + \dfrac{x-3}{4} \;=\; x - \dfrac{3x-5}{4}$

Example 3:

$4x - 1 = 4x$ has no solutions. We can see that by the fact that both sides only have one variable term, $4x$. If we subtracted $4x$ from both sides, we'd end up with the equation $-1 = 0$.

Example 4:

$6x - 2 = 2(3x - 1)$ has an infinite number of solutions. If we use the distributive property on the right side, it becomes $6x - 2$. So, both the left and right sides have an identical expression. If we subtracted $6x$ from both side and added 2 to both sides, we would end up with the equation $0 = 0$.

---

4. Tell, without solving the equations, whether each equation has one unique solution, no solutions, or an infinite number of solutions.

   a. $5x + 6 = 5x + 4$

   b. $5x + 6 = 4$

   c. $5x + 6 = 6$

   d. $5x = 4 + 5x - 4$

5. **a.** How many solutions does this equation have?

   b. Modify the equation so that it has an infinite number of solutions.

   $$2y + 4 = 7 + 2y$$

6. **a.** How many solutions does this equation have?

   b. Modify the equation so that it has no solutions.

   $$3w - 4 = 5w$$

7. Follow the instructions to make up an equation with no solutions.

   - Start with a false equation, such as 2 = 5.
   - Add $3x$ to both sides.
   - Subtract 5 from both sides.
   - Use the distributive property "backwards" and factor $3x - 3$ using 3 as the common factor.
   - Subtract $2x$ from both sides, and combine the like terms on the right side.

   $$2 = 5$$
   $$3x + 2 = 5 + 3x$$

8. Give an example of each type of equation that has $20 - 5t$ on the left side of the equation.

   a. No solutions:

   $20 - 5t =$

   b. One solution:

   $20 - 5t =$

   c. An infinite number of solutions:

   $20 - 5t =$

9. Solve both problems, and compare them.

   a. The sum of two consecutive whole numbers is 5,437. What are the numbers?

   b. The difference of two whole consecutive numbers is 1. What are the numbers?

   Hint: Let x be the first of the two consecutive whole numbers. What are the other, in terms of x?

Puzzle Corner For what value of a would the equation $ax + 9 = 8 - 5(2x + 4)$ have *no* solutions?

Percent Word Problems

> **Example 1.** What should the price of an item be, if you want its 12% off discount price to be $90?
>
> Let p be this unknown price. Decreasing it by 12% means that 88% of it is left, and 88% is 0.88.
>
> We get the simple equation $0.88p = \$90$, from which $p = \$90/0.88 \approx \$102.27$.

You may use a calculator in all the problems in this lesson.

1. In the equation $1.09p = \$108$, p signifies the price of an item. What will you find out by solving this equation?

  (i) The new price of an item that cost $108 but now is increased by 109%;

  (ii) The new price of an item that cost $108 but now is increased by 9%;

  (iii) The price of an item *before* a price increase, when the item costs $108 after a 9% price increase;

  (iv) The price of an item *before* a price increase, when the item costs $108 after a 109% price increase.

2. An airline increased its workforce by 6%, and now has 13 250 workers.
How many did they have before the increase?

3. Eggs used to cost $5.50 per dozen, but the store owner increased the price by 8%.
What is the new price?

4. What is the percent of increase if the average temperature over a certain period of time increased from 12.1°C to 12.6°C?

Example 2. A table now costs $150. How much should the store owner increase its price, so that when he has a 15% off sale, the item will sell for $150?

Note: You might get more out of this example if you try to write an equation for it yourself, first!

The unknown is how much the price is increased. Let's denote that by x. Then, the new price is $150 + x$. When this new price is discounted by 15%, it means that 85% of it is left. We want what is left to equal $150. So, the equation we get is $0.85(150 + x) = 150$.

5. Solve the equation from example 2.

6. A price is first increased by 12%, then lowered by 5%, and now it is $154.
   What was it before these changes?

7. An item had three price increases: for 2%, for 5%, and for 3%. Now it costs $62.88.
   What was its price before these increases?

8. A merchant has tables to sell that currently cost $400. He wants to increase the price by some amount so that when he sells the tables at 20% discount, the customer will pay $390. By how much should he increase the price?

9. A $92 dress is on sale for 25% off, so you buy it! A sales tax is added to your total, and you pay $72.45. What is the sales tax percentage?

Miscellaneous Problems

1. Tommy is racing with his dad. They both start at the same time, but Dad lets Tommy start 20 m ahead of him on a race track. Tommy runs at a speed of 3 m/s, and Dad runs at a speed of 6 m/s.

   a. Write an expression for Dad's position on the track at t seconds.

   b. Write an expression for Tommy's position on the track at t seconds. Take into account he started at the 20 metre-point.

   c. Now, make those two expressions equal, and solve the equation you get. What question will your solution answer?

2. A water pipe busted in your house! You are comparing the pricing of two plumbers. Antonio's PipeFix charges $45 per hour, plus a fixed fee of $35 for travelling costs. Paul the Plumber quotes you a flat fee of $220.

   a. If it is estimated that the job will take 3 ½ hours, which service is a better deal?

   b. How many hours of work does Antonio's PipeFix fee equal the same amount as Paul the Plumber?

3. To print a manual at Express Print costs $0.07 per page, plus there is a fixed fee of $2.50.
   To do the same at Sherry's Office Supplies costs $0.10 per page with no additional fees.

   a. If you are printing 500 pages, which service is a better deal?

   b. What page count do the two services equal the same cost?

4. To rent a stroller at an animal park costs $7.50 (a fixed fee) plus $2.50 per hour.

   a. Write an expression that gives the total cost of renting the stroller for m hours.

   b. Susan's bill for renting the stroller came to $25. How many hours did she use it?
   Solve this first with mental math.

   c. Now write an equation for the question in (b), and solve it.

5. The animal park raised the fees for renting a stroller. The fixed fee is now $9 and the rate per hour increased also. Susan went for 3.5 hours and paid $20.90. How much is the new rate per hour?

Puzzle Corner

Henry and Sammy work together as a team in lawn care. They charge $30 per hour for labour (per worker). But Sammy has the bad habit of never being on time and always coming ½ hour late to work.

One particular day, Henry started work at 2 PM and Sammy started ½ hour later.
At what time will their combined earnings for that day reach $100?

Hint: Write an expression for how much Henry has earned in t hours since 2 PM.
Then do the same for Sammy.

Mixed Review Chapter 3

1. Change the base of each power expression.

a. $8^6 = (2^{\boxed{}})^6 = 2^{\boxed{}}$	**b.** $64^3 = (4^{\boxed{}})^{\boxed{}} = 4^{\boxed{}}$	**c.** $100^5 = (10^{\boxed{}})^{\boxed{}} = 10^{\boxed{}}$

2. In each case, find the value of the two similar expressions.

a.	b.	c.	d.
$5 \cdot 4^3 =$	$(-5)^3 =$	$-3^{-2} =$	$2 \cdot 5^{-2} =$
$(5 \cdot 4)^3 =$	$-5^3 =$	$(-3)^{-2} =$	$(2 \cdot 5)^{-2} =$

3. The mass of the sun is about $2 \cdot 10^{30}$ kg, and the mass of the earth is about $6 \cdot 10^{24}$ kg. About how many times the mass of the earth is the mass of the sun? Give your answer in scientific notation.

4. Join equivalent expressions with a line.
Some expressions will not be joined.

$\dfrac{a^2}{b^2}$	$9a^6$	$(3a^3)^2$	$-3a^2$
$3a^6$	$\dfrac{(3a)^2}{b}$	$\left(\dfrac{a}{b}\right)^2$	$\dfrac{3a^2}{b}$
$\dfrac{9a^2}{b}$	$\dfrac{a^2}{b}$	$3a^{3^2}$	$3a^9$

5. Find the measure of angle x. Explain your reasoning.

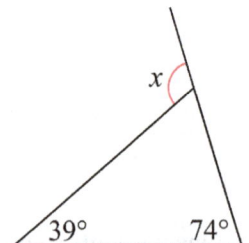

6. Find the volume of this triangular prism, if its height is 36 cm and its base is a right triangle with 10 cm, 15 cm, and 18 cm sides.

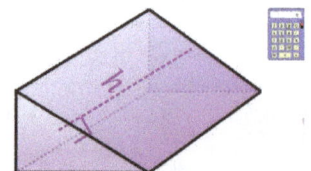

7. Two parallel lines, L_1 and L_2, are cut by a transversal.
   If angle x is 76°, find the angle measure of y, and explain
   how you know that.

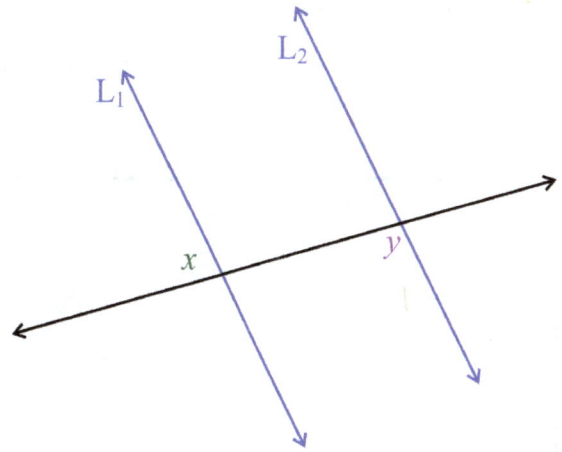

8. A triangle with vertices A(−1, 2), B(−3, 2),
   and C(−4, 4) is first reflected in the y-axis and
   then rotated 90° clockwise around the origin.
   What are the coordinates of the vertices of the
   resulting triangle?

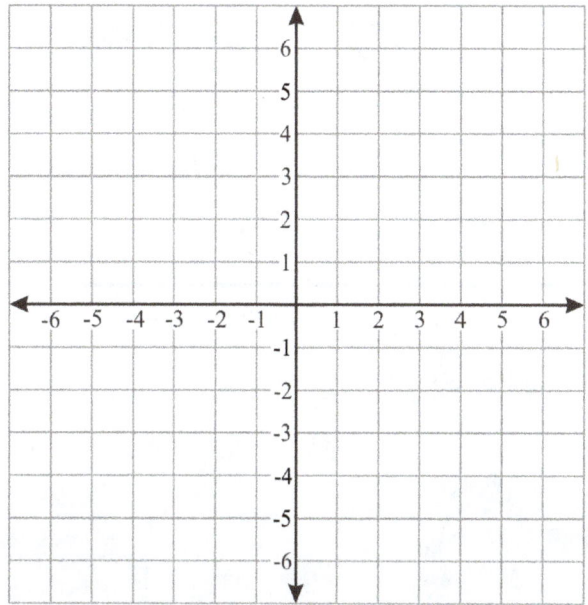

9. Dilate each figure with origin as center and with
   the given scale factor.

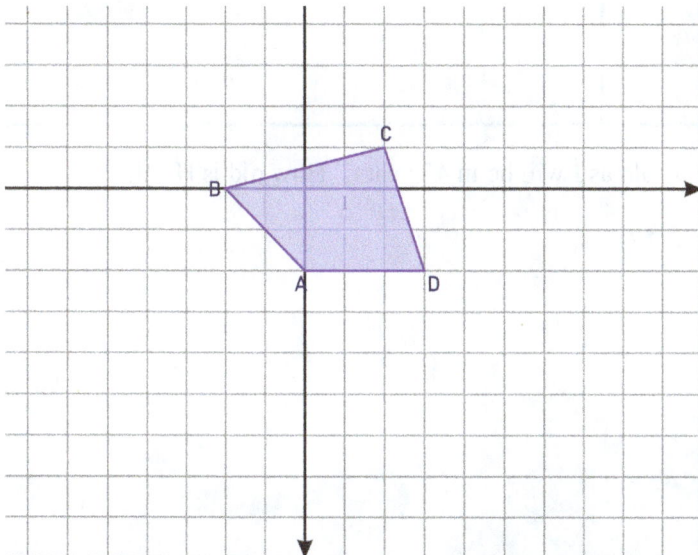

a. scale factor 3	b. scale factor 1/2

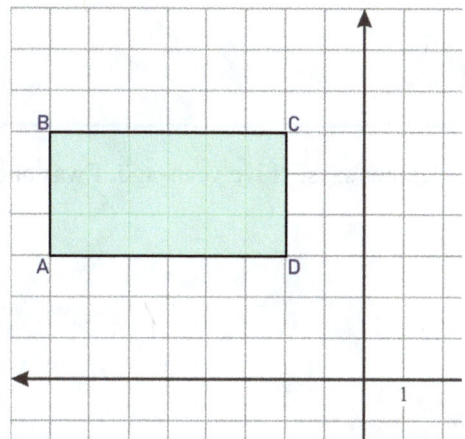

Chapter 3 Review

1. Solve.

a. $7x - 3(x - 5) - 2x = 10$	**b.** $x - \dfrac{1}{3} = \dfrac{3}{4}x - 2$
c. $20 - q = -q + 2(q - 5) - 6q$	**d.** $-52 - 2(x + 14) = 80 - 11x + x$

2. Heather says, "Five years ago, I was one-third as old as I will be in 45 years." How old is Heather?

3. Solve.

a. $\dfrac{x-1}{2} = \dfrac{3x+2}{7}$	**b.** $x - \dfrac{1-3x}{2} = 5$

4. The density of a liquid is denoted by the Greek letter ρ (rho), and is given by the formula $\rho = \dfrac{m}{V}$, where m is the mass of the liquid and V is its volume.

   a. Solve this formula for V.

   b. If the mass of a certain liquid is 4.6 kg, and its density is 850 kg/m³, what volume does it occupy? Give your answer in cubic meters *and* in cubic centimeters.

   Hint: To find the conversion factor between cubic meters and cubic centimeters, fill in:

   $1 \text{ m}^3 = 1 \text{ m} \times 1 \text{ m} \times 1 \text{ m}$

   $= 100 \text{ cm} \times \underline{\hspace{1cm}} \text{ cm} \times \underline{\hspace{1cm}} \text{ cm}$

   $= \underline{\hspace{2cm}} \text{ cm}^3$

5. Give an example of each type of equation that has $8x - 2$ on the left side of the equation.

   a. No solutions:

   $8x - 2 =$

   b. One solution:

   $8x - 2 =$

   c. An infinite number of solutions:

   $8x - 2 =$

6. Solve. Round the solutions to two decimals.

a. $\quad 0.6 + 9.4x - 2 = x - 4.8x$	b. $\quad 4 - 0.3(2y + 2.8) = 1.5(y - 3)$

7. Sarah bought a total of 16 cartons of almond milk, some for the price of $4.50 and some for $5.20. Her total came to $76.20. How many of the cheaper ones did she buy?

8. You're selling homemade bread for $8.50 per loaf. By how much should you increase the price, so that when you later on have a 25% off sale, the sale price will be $7.50?

9. The price of an item is reduced by 35%, and then a 7% sales tax is added. You pay $53.90. What was the original price of the item?

Chapter 4: Introduction to Functions
Introduction

The fourth chapter of Math Mammoth Grade 8 covers various introductory topics from the theory of functions. These topics prepare students for studying functions in great detail in high school math, and even include preparatory ideas for calculus (rate of change).

The first lesson focuses on the basic definition of a function, as a relationship between two sets that assigns exactly one output for each input. It also briefly explains the range and domain of a function, even though those terms are not required in the CCS.

Next, we study the rate of change in the context of linear functions. Students calculate the rate of change from functions given as a table of values or from their graphs. They also encounter nonlinear functions and calculate the rate of change for those in specific intervals.

Then, students learn about the initial value of a function (its value when the input is zero), and learn that the equation $y = mx + b$ defines a linear function. They write and plot equations of that form to model linear relationships. We also spend one lesson looking at linear versus nonlinear relationships.

The following major topic is describing functions. Students analyse a graph and tell whether a function is increasing, decreasing, or constant; linear or nonlinear. They sketch a graph matching a given verbal description, and interpret given graphs of nonlinear functions in a variety of the real-life contexts.

Lastly, students compare properties of two functions represented in different ways (algebraically, graphically, numerically in tables, or by verbal descriptions). For example, distance as a function of time is given as an equation for one airplane, and as a graph for another, and students answer questions concerning the speed and distance of the two airplanes.

Pacing Suggestion for Chapter 4

This table does not include the chapter test as it is found in a different book (or file).
Please add one day to the pacing if you use the test.

The Lessons in Chapter 4	page	span	suggested pacing	your pacing
Functions	191	*4 pages*	1 day	
Linear Functions and the Rate of Change 1	195	*4 pages*	1 day	
Linear Functions and the Rate of Change 2	199	*3 pages*	1 day	
Linear Functions as Equations	202	*3 pages*	1 day	
Linear versus Nonlinear Functions	205	*3 pages*	1 day	
Modelling Linear Relationships	208	*4 pages*	1 day	
Describing Functions 1	212	*3 pages*	1 day	
Describing Functions 2	215	*3 pages*	1 day	
Describing Functions 3	218	*4 pages*	1 day	
Comparing Functions 1	222	*3 pages*	1 day	
Comparing Functions 2	225	*2 pages*	1 day	
Chapter 4 Mixed Review	227	*3 pages*	1 day	
Chapter 4 Review	230	*4 pages*	2 days	
Chapter 4 Test (optional)				
TOTALS		*43 pages*	14 days	

Helpful Resources on the Internet

We have compiled a list of Internet resources that match the topics in this chapter, including pages that offer:

- **online practice** for concepts;
- online **games**, or occasionally, printable games;
- **animations** and interactive **illustrations** of math concepts;
- **articles** that teach a math concept.

We heartily recommend you take a look! Many of our customers love using these resources to supplement the bookwork. You can use these resources as you see fit for extra practice, to illustrate a concept better and even just for some fun. Enjoy!

https://l.mathmammoth.com/gr8ch4

Scan me

Functions

A **function** is a rule or a relationship between two sets that assigns **exactly one output for each input.** We also use the word **mapping** for a function.

Example 1. The illustration below shows a simple function that maps each animal to its favourite sleeping place.

Each animal has a sleeping place, and only one, so this is a function.

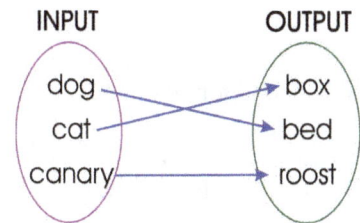

Example 2. The table lists the name of seven children, and the month when each child has their birthday. Notice that several of them have their birthday in December. Is this a function?

Input	Allie	Julie	Danny	Juan	Pete	Bob	Samantha
Output	September	December	December	June	August	December	February

Yes. The definition only requires that there has to be exactly one output for each input; **the outputs don't have to be unique.**

1. The relationship shown on the right is *not* a function. Why?

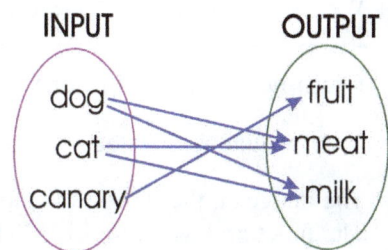

2. A function machine "ingests" a number (the input) and "spits out" another (the output) based on some rule. This function machine turns any number n into $4n + 1$.

   a. Number −7 is just going in. What will be the output?

   b. Number 17 just came out. What was the input?

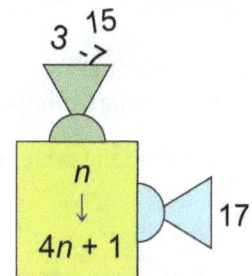

3. Potatoes cost $3 per kilogram. Fill in the tables #1 and #2.

   Does each table represent a function? Explain.

#1 (Input) Weight	#1 (Output) Cost
1 kg	$3
2 kg	
3 kg	
5 kg	
12 kg	

#2 (Input) Cost	#2 (Output) Weight
$12	
$30	
$48	
$72	
$90	

191

4. The table lists seven children, and each child's favourite colour.

Input	Allie	Julie	Danny	Juan	Pete	Bob	Samantha
Output	pink and blue	blue	grey	yellow	blue and red	?	purple

Is this a function?   If not, change it in some manner(s) so it *is* a function.

5. T is a function that maps the name of a month to the number of days in it.

   a. Create a depiction of T using
   a diagram like in example 1.

   b. If you reverse the inputs and outputs, is the
   resulting relationship a function? Explain.

If the inputs and outputs are numbers, we can plot **a graph of the function** in the coordinate grid. Each input-output pair is viewed as an ordered pair (a single point).

We also use the terms "independent variable" for the input, and "dependent variable" for the output.

Example 3. Let F be the function $(1, 2), (3, 0), (5, 3), (7, 1)$.

Note: A function *can* be given as a list of ordered pairs.

The image on the right is the plot of F; yet the plot is *not* F. The function F is the specific list of inputs and outputs, or the relationship itself.

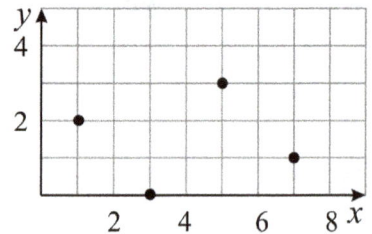

---

6. Let G be the function that maps each integer from −4 to 4 to its square minus one.

   a. Fill in the table, listing the ordered pairs of G.

Input (*x*)	−4	−3	−2						
Output (*y*)	15								

   b. Make a plot of G.

   c. If you reversed the inputs and the outputs, would the
   relationship still be a function? Explain.

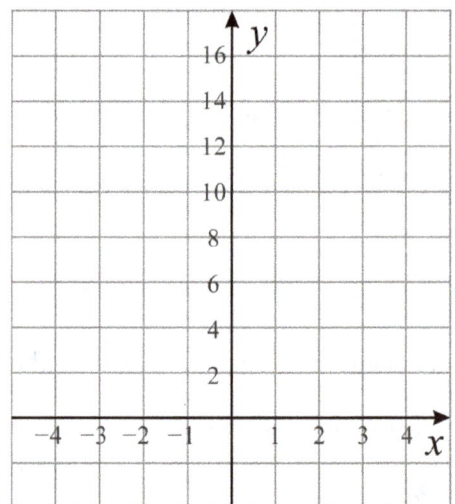

Example 4. Mary bicycled from her home to a friend's house. The table shows the distance (d) Mary had covered at specific amounts of time (t).

Input (t)	5 min	10 min	12 min	13 min	15 min	18 min	20 min	22 min
Output (d)	0.8 km	1.5 km	1.9 km	1.9 km	1.9 km	2.4 km	2.7 km	3 km

We say that **distance is a function of time.** The output variable, or the dependent variable, is always said to be a function of the input (or independent) variable. This means that for each moment of time (input) there is a specific distance she has travelled (output).

Is it true in reverse? Is *time* a function of *distance*?

This means we consider distance as the input, and time as the output. If yes, then for each distance (input), there is exactly one time (output). Is that so in this case?

7. Is Age a function of Name? Explain.

Is Name a function of Age? Explain.

Name	Age
FenFen	14
Larry	15
Pierre	13
Sam	12
Amy	14

Age	Name
14	FenFen
15	Larry
13	Pierre
12	Sam
14	Amy

8. Choose the relationships that are functions.

(1)
Rainfall (mm)	2	0	0	5	0	13	0
Day of month	6	7	8	9	10	11	12

(2) Let S be a rule that takes any number x as input, and gives $4x + 1$ as output.

(3) Input is a zip code, output is a person that lives there.

(4) Input is a person's first name, output is their bank account number.

9. Plot the following points that give the age (in years) and the height (in metres) of various children.

(2, 0.8) (5, 1.05) (10, 1.40) (9, 1.31) (6, 1.17) (5, 1.09)

a. Is this a function? Explain.

b. What is the independent variable? The dependent variable?

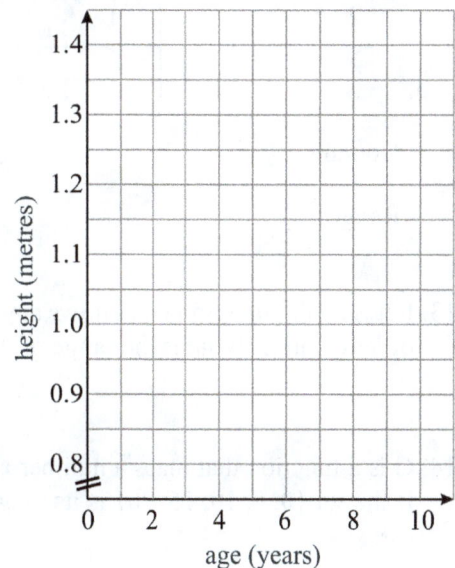

The **domain** of a function is the set of inputs. The **range** of a function is the set of outputs.

Let's go back to example 3, where we had kindergartners and their birthday months.

Input	Allie	Julie	Danny	Juan	Pete	Bob	Samantha
Output	September	December	December	June	August	December	February

The domain of this function is the list of the children's names. To write it as a set, we enclose the items of the set in curly brackets: {Allie, Julie, Danny, Juan, Pete, Bob, Samantha}.

The range of this function is {September, December, June, August, February}.

10. **a.** Change some thing(s) in this table so it is a function.

   b. Give the domain of the function.

   c. Give the range of the function.

Input	Output
Name	**Grade level**
Jenny	8
Pedro	7
Ann	8
Marsha	
Rob	9
Ann	6

11. Let F be the function that maps a number x to $2x + 1$.
    Let the set {0, 1, 2, 3, 4, 5} be its domain.
    What is its range?

12. Give the domain and range of each function.

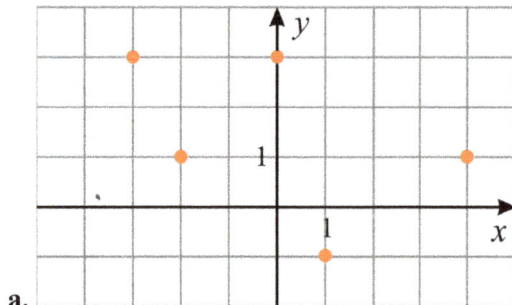

a.

Domain:

Range:

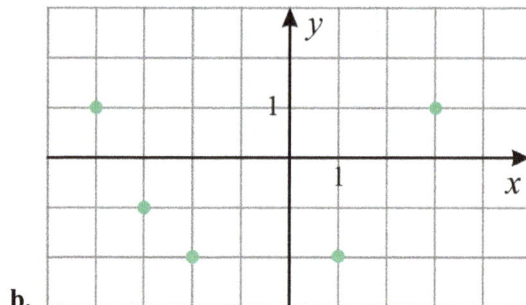

b.

Domain:

Range:

13. Let S be the function that allows any word from this sentence as the input, and the output is the number of letters in it. What is the range of this function?

14. G is a function that maps a number x to $x - 5$.
    If the set {0, 5, 10, 15, 20} is its range, what is its domain?

Linear Functions and the Rate of Change 1

If the graph of a function consists of points that fall on a single line, it is a **linear function**.

We will define a linear function in a different manner later, but for now, this is sufficient, so let's look at some examples.

Example 1. The input and output values in the table below define a function. Notice the patterns: the x-values increase by ones, and the y-values increase by 3s.

Input (x)	0	1	2	3	4	5	6
Output (y)	3	6	9	12	15	18	21

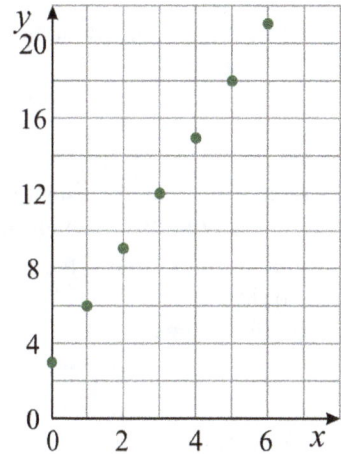

The graph shows that the points fall on a line. This is a linear function.

The **rate of change** of a function is the rate at which the output values change as compared to the change in the input values.

We calculate it as the ratio of $\dfrac{\text{change in output values}}{\text{change in input values}}$.

In the context of this graph, **rate of change** = $\dfrac{\text{difference in } y\text{-values}}{\text{difference in } x\text{-values}}$.

In this case, each time the x-values increase by 1, the y-values increase by 3. **The rate of change is 3/1 = _3_ .**

Example 2. The price of bananas is a function of their weight. What is the rate of change?

Weight in kg (input)	0	2	5	10	12	15
Price in $ (output)	0	5	12.50	25	30	37.50

Check how much the output (price) changes for a certain change in the input (the weight). For example, when the weight increases from 0 to 2 kg, the price increases from $0 to $5, or by $5. This happens also when the weight increases from 10 to 12 kg: the price increases $5 (from $25 to $30).

$$\text{Rate of change} = \frac{\$5}{2 \text{ kg}} = \$2.50/\text{kg}$$

Note that if the independent and dependent variables have units, **we include the units in the rate of change**.

This rate of change tells us that for each one-kilogram increase in weight, the price increases by $2.50.

1. **a.** Calculate the rate of change in example 2, using the increase in weight from 5 to 10 kg, and the corresponding increase in price. Do you get the same rate of change as calculated in the example?

   b. Do the same using the input values 10 kg and 15 kg.

2. What is the rate of change? Don't forget the units!

a.

Input (*t*)	2 hrs	3 hrs	4 hrs	5 hrs	6 hrs	7 hrs
Output (*d*)	$30	$45	$60	$75	$90	$105

b.

Input (*t*)	2 L	4 L	6 L	8 L
Output (*d*)	2.8 kg	5.6 kg	8.4 kg	11.2 kg

3. If a linear function contains the points (4, 15) and (9, 18), what is the rate of change?

4. A train travels at a constant speed, travelling 40 km in 20 minutes. Function D gives the distance (*d*) in kilometres that the train has travelled in *t* hours.

a. Fill in the output values.

t (hours)	0 hrs	1 hr	2 hrs	3 hrs	4 hrs	5 hrs	6 hrs
d (km)							

b. What is the rate of change?
   Use hours and kilometres.

5. Mr. Stevenson, a gardener, is being paid a base salary of $400 per week for taking basic care of the grounds at a college, plus $25 per hour for certain special tasks. We can model his weekly earnings (E) with the function E = 400 + 25*t* where *t* is the number of hours he works at the special tasks.

a. How much does he get paid if he works five hours at the special tasks in a week?

b. How many hours would he need to work at the special tasks to earn $575 in a week?

c. What is the rate of change of this function?

6. Function D has the rate of change of (7 metres)/(20 minutes), and at 0 minutes, the output value is 0.5 metres.

a. Fill in the table.

Input (t)	0 min	10 min	20 min	30 min	40 min	50 min	60 min
Output (d)	0.5 m						

b. What could this depict?

7. The price of potatoes increases by $10 each time the weight increases by 5 kg.
   How do the the rate of change and unit price compare in this situation?

Example 3. The graph shows a plot of a function. To determine the rate of change from a graph, we look at the coordinates of *two* points.

$$\text{Rate of change} = \frac{\text{difference in their } y\text{-values}}{\text{difference in their } x\text{-values}}$$

If the function is linear, you can look at *any two* points in the graph in order to determine the rate of change. Here, we use the points (1, 5) and (2, 3).

As the *x*-values **increase by one** (from 1 to 2), the *y*-values **decrease** by two (from 5 to 3). This means the difference in the *y*-values is −2.

The rate of change is negative, and is $\frac{-2}{1} = -2$.

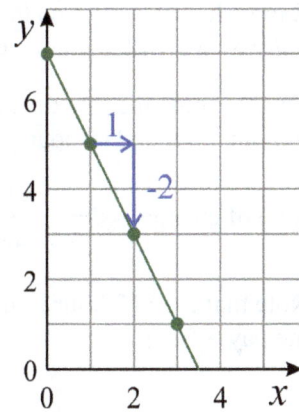

8. Find the rate of change for each function. Note that it can be negative, and/or a fraction.

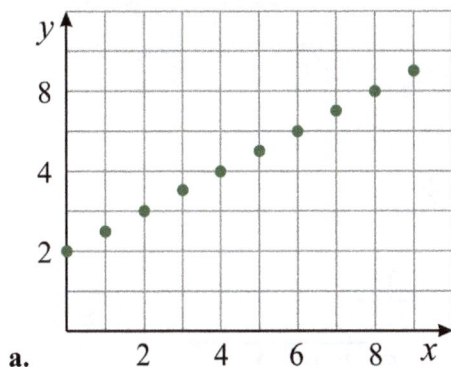

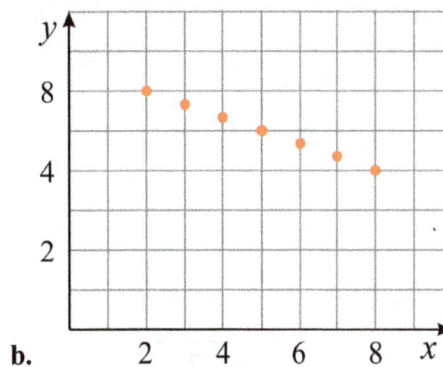

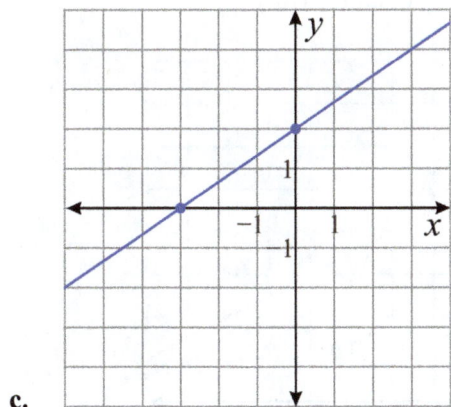

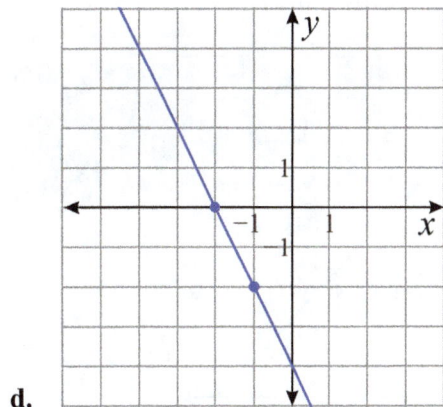

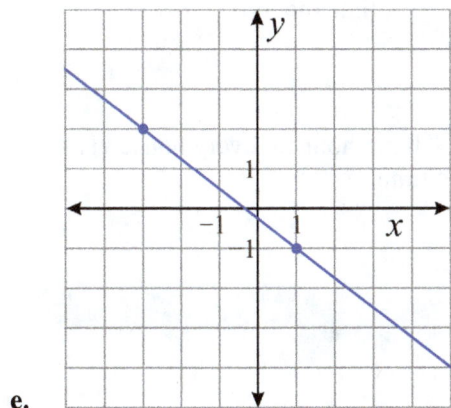

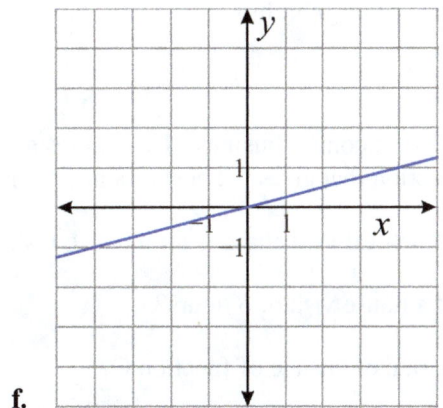

a.

b.

c.

d.

e.

f.

197

Example 4. The plot on the right includes the units *hours* and *dollars*. We will need to include those in the rate of change.

To determine the rate of change, we can use the two points (0 hours, $0) and (6 hours, $90).

Rate of change = $\dfrac{\$90 - \$0}{6 \text{ hours} - 0 \text{ hours}} = \dfrac{\$90}{6 \text{ hours}} = \$15/\text{hour}$

Note that the $15/hour is also the <u>unit rate</u> (when time = 1 hour, the pay = $15).

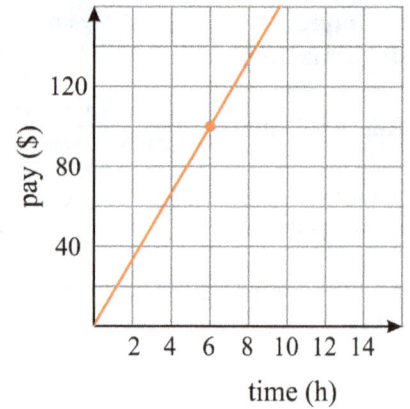

9. Find the rate of change. Include the units.

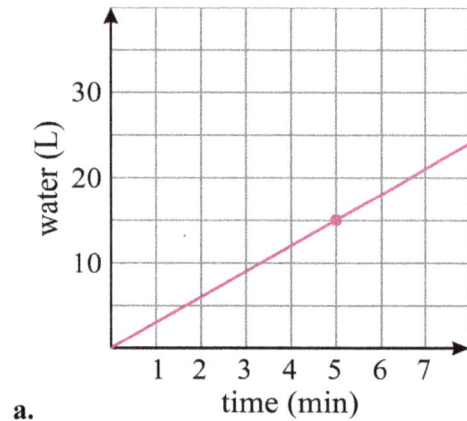

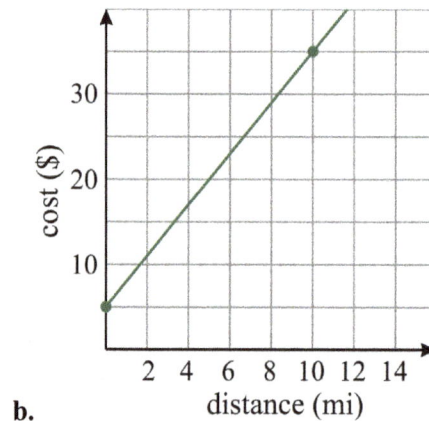

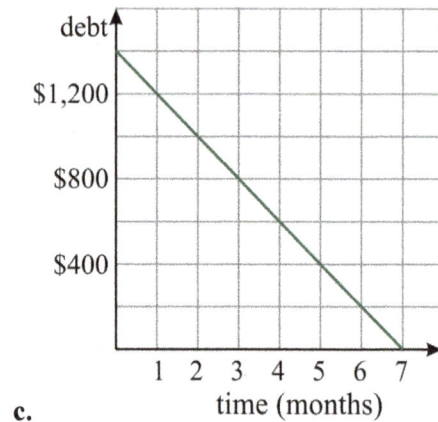

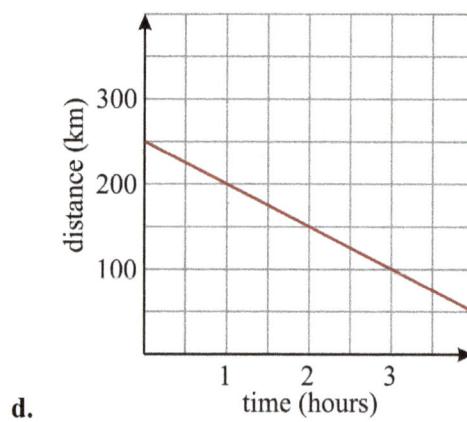

a.

b.

c.

d.

10. Edward runs a pest control business. He charges a $80 base fee plus $50 per hour for every house visit. Let C be a function that gives the cost of a house visit as a function of time.

  a. Find how much Edward charges for a house visit lasting 3 hours.

  b. How about a house visit of 5 hours?

  c. What is the rate of change of function C?

Linear Functions and the Rate of Change 2

Example 1. Susan went for a jog. The table shows the distance she has covered as a function of time.

Time (min)	0	10	20	30	40	50	60	70
Distance (km)	0	1.5	3	4.5	5.25	6	6.75	7.5

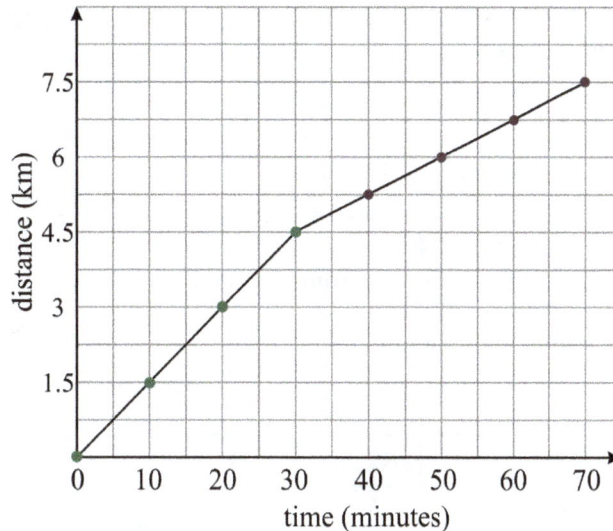

Is this a linear function? (Is it even a function?)

It *is* a function, but *not* a linear function. The points do not fall on a *single* line.

We can also notice this when examining the rate of change.

If we use the points (0, 0) and (20, 3), we will get the rate of change as $\dfrac{3 \text{ km}}{20 \text{ min}} = 0.15$ km/min.

But if we use the points (50, 6) and (70, 7.5), we will get the rate of change as $\dfrac{1.5 \text{ km}}{20 \text{ min}} = 0.075$ km/min.

There are two different rates of change or **speeds** going on: from 0 till 30 minutes, she is jogging at the rate of 0.15 km per minute. From 30 through 70 minutes, her speed is 0.075 km per minute.

How can you see the two different rates of change in the graph?

Since the rate of change is not constant, but varies, the function is *not* linear.

This fact actually gives us another definition for a linear function:

If the rate of change of a function is constant, the function is linear, and otherwise not.

---

1. If we limit the variable *time* to be from 30 to 70 minutes only, is the function in Example 1 linear? Why or why not?

2. Calculate Susan's average speed between 20 and 40 minutes.

3. The graph shows a plot of a certain function.

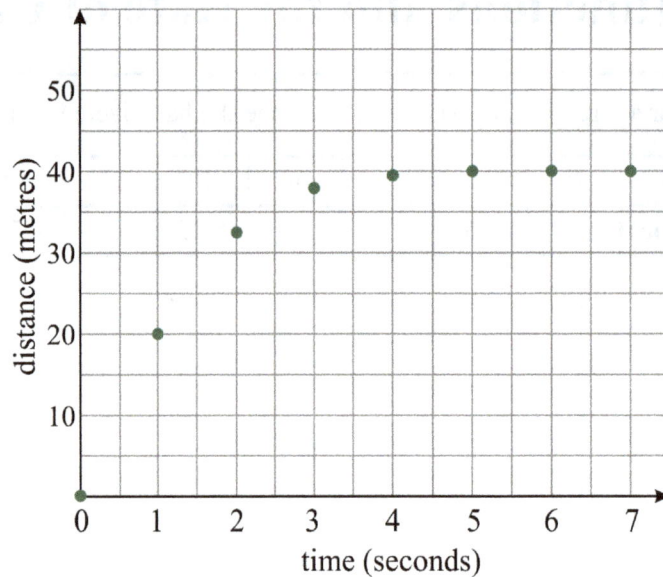

a. Fill in the table with the coordinates of the points.

Point	A	B	C	D	E	F	G	H
Time (sec)	0	1	2	3	4	5	6	7
Distance (m)	0							

What is the rate of change (or speed)...

b. from point A to point B?

c. from point B to point C?

d. from point F to point G?

e. Is this a linear function? Explain.

f. What situation in real life could produce this data?

4. The ordered pairs below define a function.

Input (x)	−5	−4	−3	−2	−1	0	1	2	3	4	5
Output (y)	8	6	4	2	0	2	4	6	8	10	12

a. Is it a linear function?   Why or why not?

b. Is there any interval or range of x-values where the function *is* linear?

5. The formula $T = 20° + 0.5t$ represents temperature (T, in Celsius degrees) as a function of time (t, in hours).

a. Fill in the table of values.

Input (*t*)	0	0.5	1	1.5	2	2.5	3	3.5	4
Output (T)									

b. Make a plot of the function. Note that you need to design the scaling for the *t*-axis.
   (The ⚡ symbol on the vertical axis signifies that the scaling does not start from zero.)

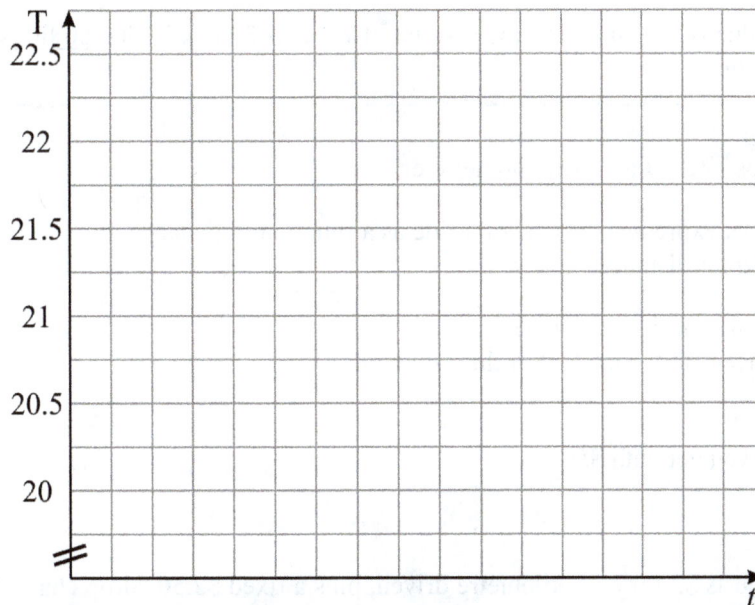

c. Is the function linear? Explain.

d. What is its rate of change between $t = 2$ hours and $t = 5$ hours?

8. Which of the two functions on the right is linear?

How can you tell?

a.

Input (*x*)	Output (*y*)
−2	17
−1	14
0	11
1	8
2	5
3	2

b.

Input (*x*)	Output (*y*)
−2	35
−1	30
0	26
1	24
2	19
3	14

Linear Functions as Equations

The **initial value** of a function is the output value when the input value is zero.

Example 1. The table below shows the volume of water (V) in a water tank as a function of time *(t)*, as the tank is being filled.

t *(minutes)*	0	1	2	3	4	5	6
V *(litres)*	80	90	100	110	120	130	140

When the input value is $t = 0$ minutes, the amount of water is 80 litres. That <u>80 litres</u> is the initial value of this function.

1. A taxi ride with Joe costs $2.50 for each kilometre driven.

   a. Write an equation that expresses the cost of a ride as a function of distance.
   Use C for cost, and *d* for distance.

   b. What is the initial value of this function?
   Explain what this means in the context of this situation.

   c. How long a ride do you get with $50?

2. A taxi ride with Jack costs $2 for each kilometre driven, plus a fixed $2.50 "drop charge" applied as soon as you get in the taxi and accept the ride.

   a. Write an equation that expresses the cost of a ride as a function of distance.

   b. What is the initial value of this function?

   c. How long a ride do you get with $50?

3. See the table below for the cost (C) of taxi rides with Andrew, as a function of distance *(d)*.

d (km)	0	2	4	6	8	10	20	30
C ($)	5	8	11	14	17	20	35	50

   a. What is the initial value?

   b. What is the rate of change?

   c. What rule determines how the cost is figured? Write the rule as an equation.

4. Comparing Joe's, Jack's, and Andrew's taxi services, which of them has the largest rate of change?

   Which is the best deal for a 10-km ride?

The equation $y = mx + b$ defines a linear function where:

- x is the independent variable (input)
- y is the dependent variable (output)
- b is the initial value, and
- m is the rate of change.

If a function is linear, we can always write an equation for it in this format.

Example 2. Marilyn is taking part in a multi-day horse race. After two days, she has travelled 260 km. On the third day, Marilyn rides at a constant speed of 15 km per hour.

The total distance she has covered in the race is a function of time, and we can write an equation for it. Our variables are time (t) and distance (d). The equation is $d = 15t + 260$.

Notice this equation is in the format $y = mx + b$. The rate of change, m, is the speed (15 km/h). The initial value, b, is 260 km.

We could also write the equation as $d = 260 + 15t$, which emphasizes the starting (initial) value, and how each hour of riding adds 15 more kilometres to it.

Since this is a *linear* function, its graph is a **line**. Therefore, to plot the function, it is sufficient to plot two points and to draw a line through them. Here, we used the points (0 hours, 260 km) and (4 hours, 320 km).

(The ⫻ symbol on the vertical axis signifies that the scaling does not start from zero.)

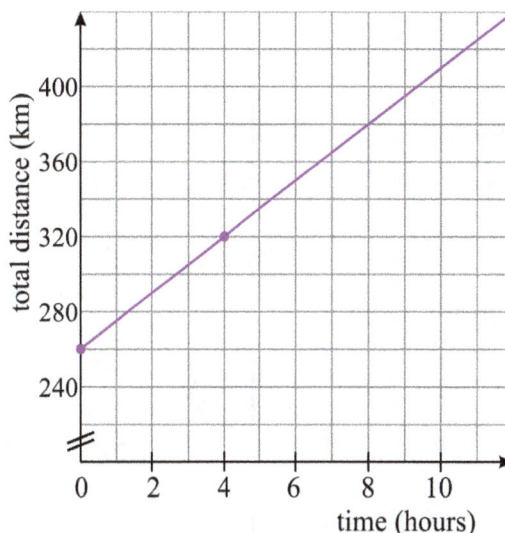

5. Carl is saving money at a steady rate. The equation $M = 50w + 32$ describes the number of dollars he has, as a function of time, where w is the number of weeks since January 1.

   a. Is this a linear function?

   b. What is the rate of change? Include the units.

   c. What is the initial value?

   d. When will Carl have at least $600?

6. **a.** Determine the rate of change and the initial value of this linear function from its plot. Don't forget the units.

   Initial value: _____

   Rate of change: _____

   b. Write an equation for the function.

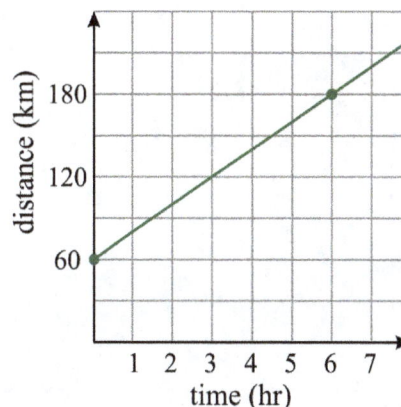

7. Going back to example 1 in this lesson, the volume of water in a tank was a function of time *(t)*.

t (minutes)	0	1	2	3	4	5	6
V (litres)	80	90	100	110	120	130	140

   a. What is the rate of change?
   What does it mean in this situation?

   b. Write an equation that gives you the volume of water as a function of time.

   c. Plot the equation you wrote in (b).

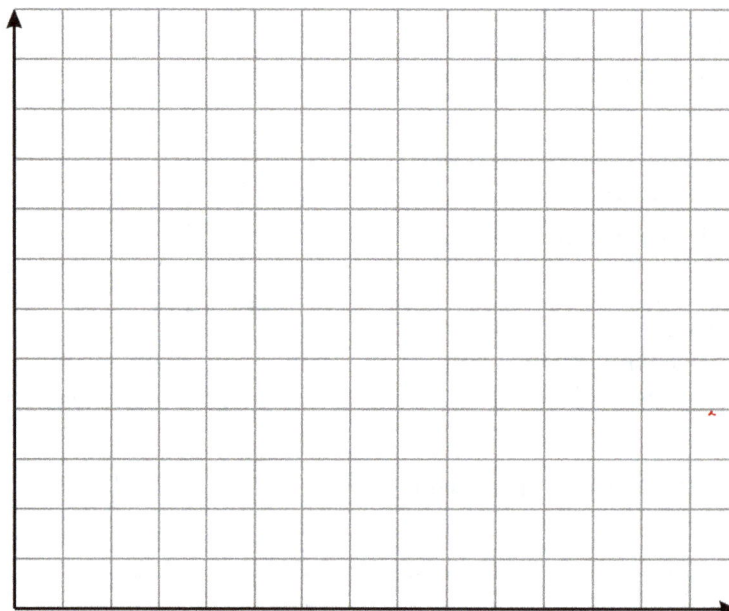

   d. How much water is in the tank after 20 minutes?

   e. When will the tank have 400 litres of water in it?

Puzzle Corner

Function F is linear, and includes the points (−9, −7) and (−4, −1).
Function G is also linear, and includes the points (−8, −9) and (−4, −3).
Which function has a larger rate of change?

Linear Versus Nonlinear Functions

Recall these facts that are true for a **linear function**:

- Its **rate of change is constant** (does not vary).
- The input & output values as ordered pairs **form a line**, when plotted in the coordinate grid.
- Any linear function can be represented with an equation of the form $y = mx + b$, where m is the rate of change and b is the initial value.

These facts give us several tools for checking whether a function is linear.

1. The graph below shows the air temperature on a certain day.

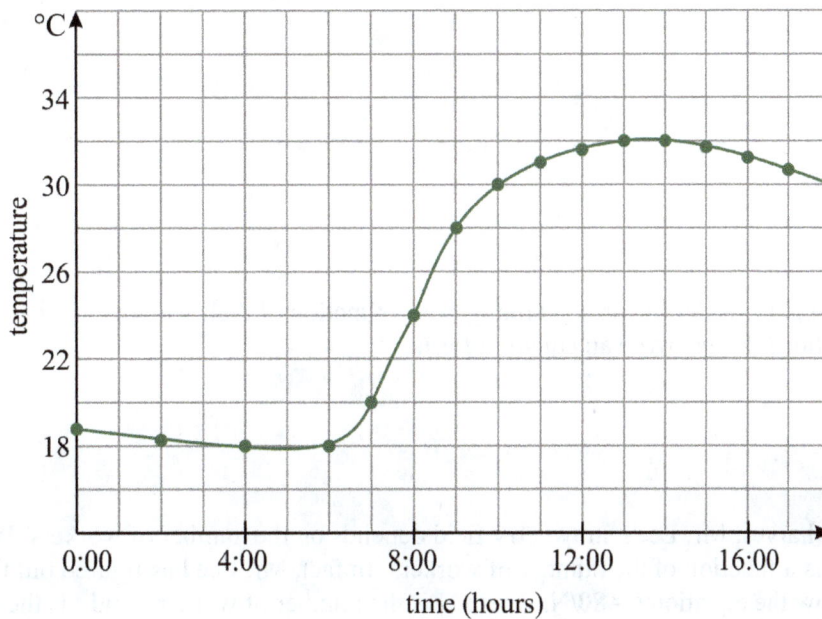

a. Is this a function? Explain.

b. Is it a linear function? How can you tell?

What is the rate of change...

c. between 0:00 and 4:00?

d. between 7:00 and 10:00?

e. Identify two time periods when the rate of change is practically zero.

f. Identify two time periods when this function is decreasing (the temperature is dropping).

g. When is the temperature rising the quickest?

2. Make up two functions for the cost of renting a bicycle as a function of time, one linear, and the other nonlinear. Give your functions as a table of values.

Function 1:

time (hours)	0	1	2	3	4	5	6	7	8	9	10
Cost ($)											

Function 2:

time (hours)	0	1	2	3	4	5	6	7	8	9	10
Cost ($)											

3. The equation $A = s^2$ gives the area of a square as a function of its side length (s). Prove that it is *not* a linear function.

4. Mangos cost $2.50/kg. Consider the price of mangos as a function of their weight. Is this a linear function? If yes, write an equation for it.

5. The time it takes to harvest Mr. Lee's strawberry field depends on the number of workers. We can say that the time to harvest is a function of the number of workers. In fact, Mr. Lee has figured out that it seems to approximately follow the equation $t = 80/N$, where N is the number of workers, and t is the time in hours.

   a. Compare the time it takes if there are 10 workers versus if there are 5 workers.

   b. Is this a linear function? Explain.

6. Give an example of a nonlinear function as an equation. (Do not use the exact equations from this lesson.)

7. Marsha goes for a jog, but she might also walk or stop for a while.

Make up a function for the total distance Marsha has travelled, as a function of time, so that the function is not linear, yet is reasonable (could happen in reality). Fill in a table of values and make a plot.

Note: An average walking speed is from 4 to 6 km/h. An average jogging speed is from 6 to 9 km/h.

time (minutes)	0	5	10	15	20	25	30	35	40	45	50	55	60
Distance (km)													

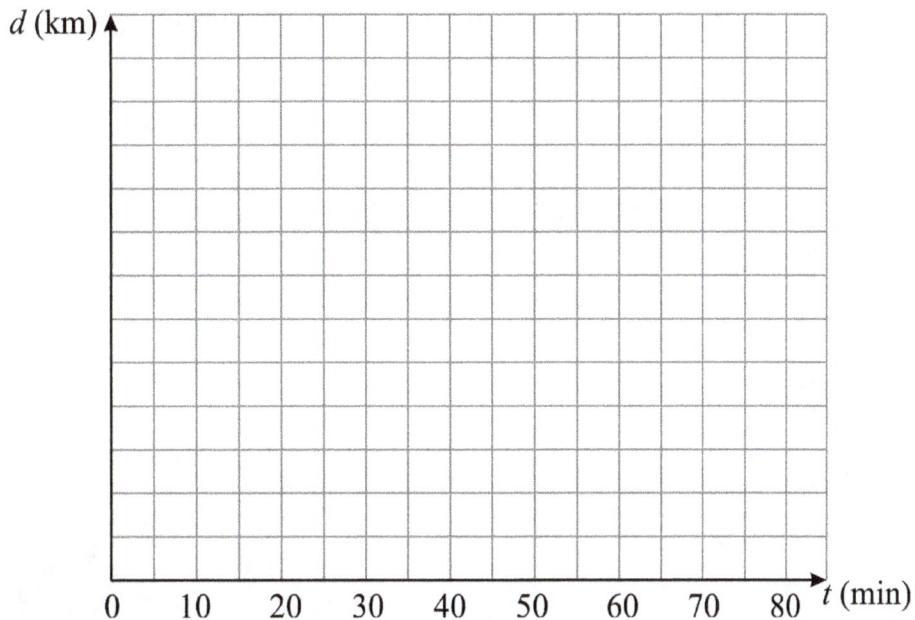

Puzzle Corner

Match each situation with a graph.
One graph will not be matched.

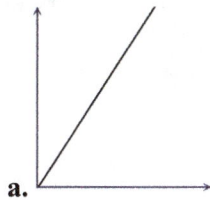

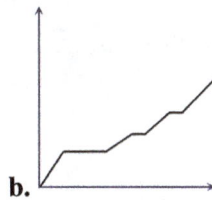

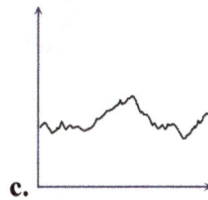

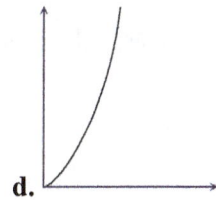

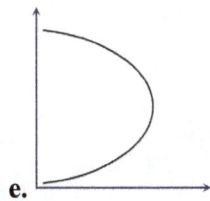

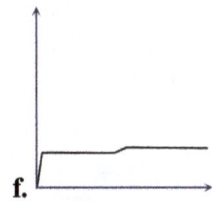

a.   b.   c.   d.

e.   f.

(i) The total distance Henry has covered on his walk with the dogs.

(ii) The cost of potatoes as per their weight.

(iii) The unemployment rate over time.

(iv) Surface area of a cube as a function of edge length.

(v) The amount of water in a swimming pool over time.

Modelling Linear Relationships

> Reminder: The equation $y = mx + b$ defines a linear function where:
>
> - x is the independent variable (input)
> - y is the dependent variable (output)
> - b is the initial value, and
> - m is the rate of change.
>
> If a function is linear, we can always write an equation for it in this format.

1. The table shows the cost of renting a specialty car for a day. The base cost is $120, with a 300-km allowance. Any distance you drive over that 300 km, you will pay an additional fee.

Km over allowance	0	20	40	60	80	100	120	140	160	180
Cost	$120	$124	$128	$132	$136	$140	$144	$148	$152	$156

   a. What is the rate of change?

   What does it signify in this situation?

   b. What is the initial value?

   What does it signify in this situation?

   c. Write an equation to model this relationship.

   d. How much is the rental cost if you drive a total of 481 km?

   e. How many kilometres over the allowance can you drive with $170?

2. Andrew has already saved $140, and from now on, he will be putting $25 into his savings every week.

   a. Write an equation for the total savings he has, as a function of time.

   b. When will Andrew have saved $490?

3. The plot shows the amount of water in a water tank as a function of time.

   a. What is the rate of change?

   What does it signify in this situation?

   b. What is the initial value?

   What does it signify in this situation?

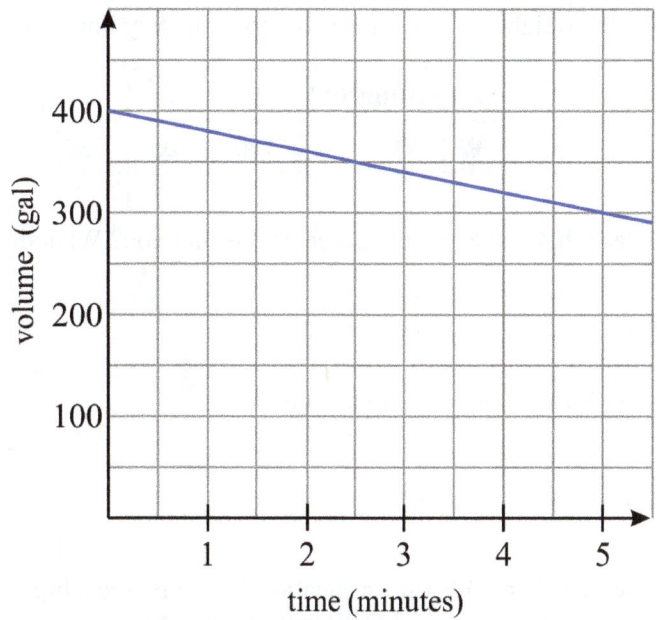

   c. Write an equation to model this relationship.

4. Sarah took a $6500 loan, and she is paying it back at the rate of $350 each month.

   a. Write an equation to represent the amount of debt (D) she has, as a function of time (t, in months).

   b. Plot your equation. Make sure that the point corresponding to *time* = 12 months fits in it.

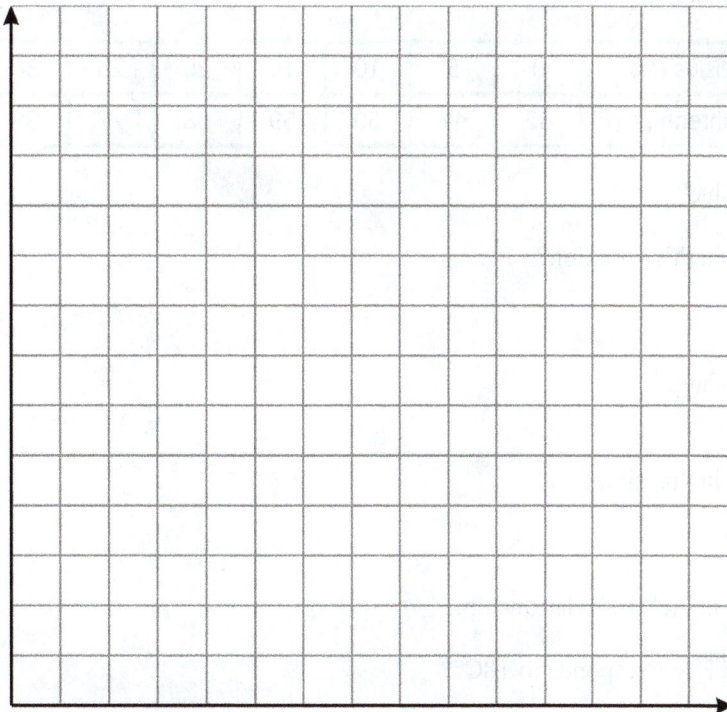

5. The weight gain of a heifer is approximately linear between the ages of 6 and 14 months.

Age (months)	6	7	8	9	10	11	12	13	14
W (kg)	180	205	230	255	280	305	330	355	380

a. What is the rate of change of this function? What does it signify in this situation?

b. Make a plot of this function.

c. (Challenge) Write an equation for the relationship between the weight (W) and the age (A).

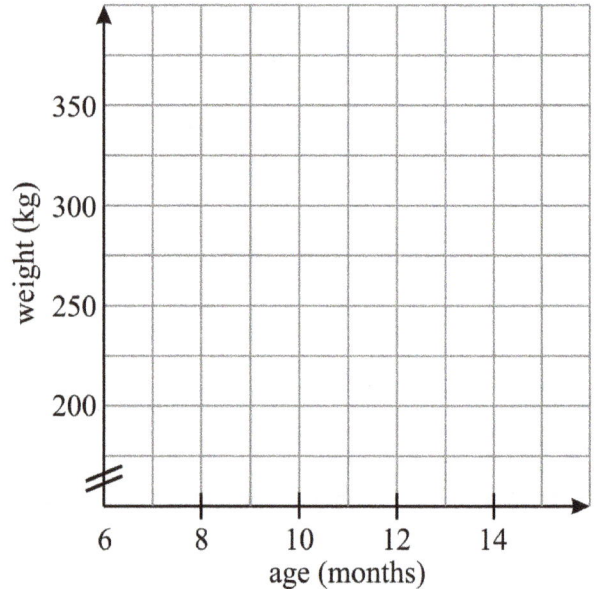

6. The table below shows how Celsius and Fahrenheit degrees correspond at different temperatures. Consider the linear function that inputs the temperature in Celsius (C), and outputs the temperature in Fahrenheit (F).

Celsius (C)	0	5	10	15	20	25	30
Fahrenheit (F)	32	41	50	59	68	77	86

a. What is the initial value?

What does it signify in this situation?

b. What is the rate of change?

What does it signify in this situation?

c. Write an equation to model this relationship.

d. What temperature in F° corresponds to 18C°?

e. What temperature in C° corresponds to 100F°?

7. The graph shows the volume of a cylindrical drinking glass with a 7-cm diameter, as a function of its height. In other words, the diameter of the glass is fixed or decided, but the height is not, and we're looking at how the volume changes as the height changes.

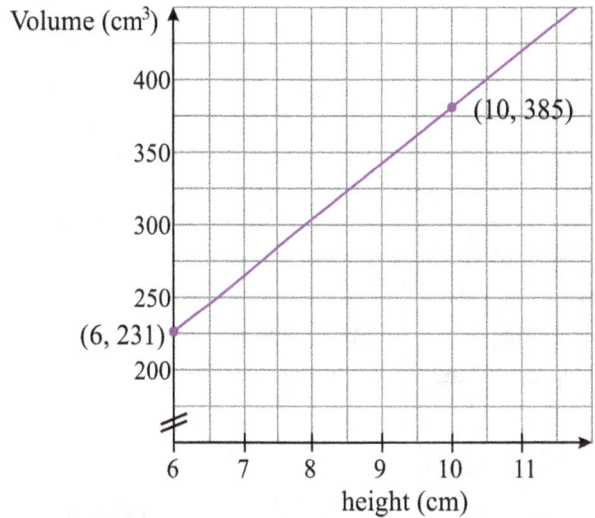

a. What is the initial value? (It is not shown on the graph, but you can figure it out with common sense.)

b. Write an equation to model this relationship, using the given points.

c. How tall would the glass need to be in order to have a volume of 500 cm^3?

d. What is the volume of the glass if its height is 7.4 cm?

e. Now use your knowledge of geometry. Write a formula for the volume of this glass. Compare it to the equation you wrote in (c). (Note: the reason they are slightly different is because the numbers in the graph above are rounded.)

Natalie is deciding between two different cars. The first one, Car 1, costs $35 000, and she would pay for it in monthly payments of $1500. Car 2 costs Y dollars, and she would pay for it in monthly payments of $1800.

Puzzle Corner

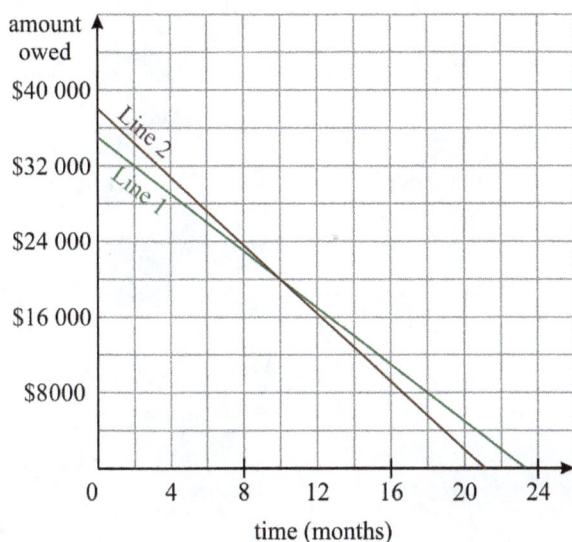

a. Match each line in the picture with the correct car/payment plan.

b. The two plans meet at (10 months, $20 000). Write an equation for the line for Car 2.

211

Describing Functions 1

A function is said to be **increasing** if its graph is continually going upwards. If its graph is continually going downwards, it is **decreasing**. A function can also be **constant** — its graph is a horizontal line.

We typically use these terms to describe a function in a certain interval. We use the notation $[a, b]$ to denote an interval from a to b.

Example 1. From the graph we can see that this function is <u>decreasing</u> in the interval $[1, 3]$ (in other words, from $x = 1$ to $x = 3$).

It is <u>increasing</u> in the interval $[3, 5]$.

In the interval $[0, 5]$ it is neither increasing nor decreasing.

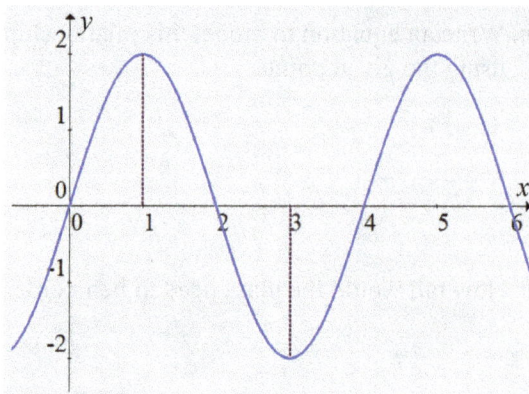

Example 2. This function is linear and increasing in the interval $[-3, -1]$.

Next, it is constant in the interval $[-1, 1.1]$. (Its output value, or y-value, is one, all through that interval.)

Then it is nonlinear and decreasing in the interval $[1.1, 1.8]$. Lastly, it is nonlinear and increasing from $x = 1.8$ to $x = 5$.

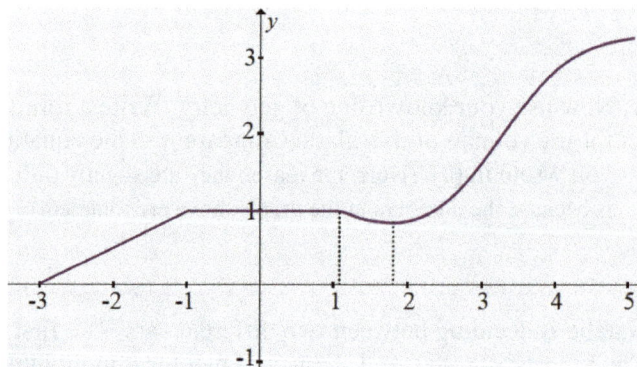

1. Describe this function by intervals where it is increasing, decreasing, or constant. Include also whether it is linear or nonlinear in those intervals.

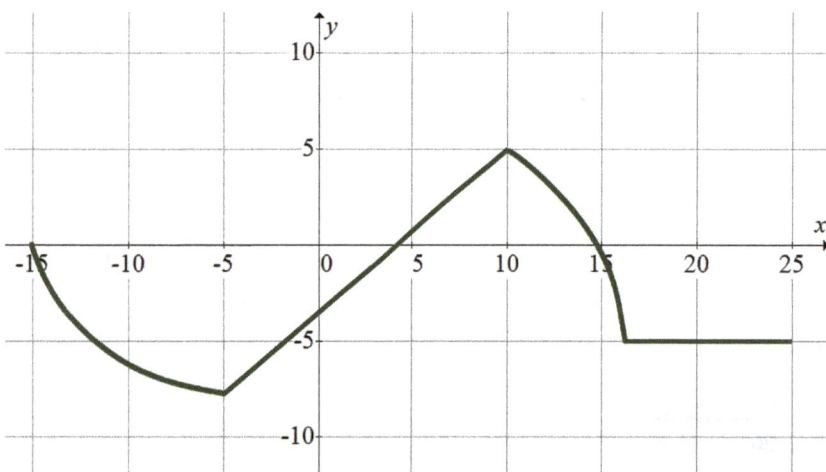

2. The graph depicts the cost of solar systems per watt sold by Henry's Sun Power store.

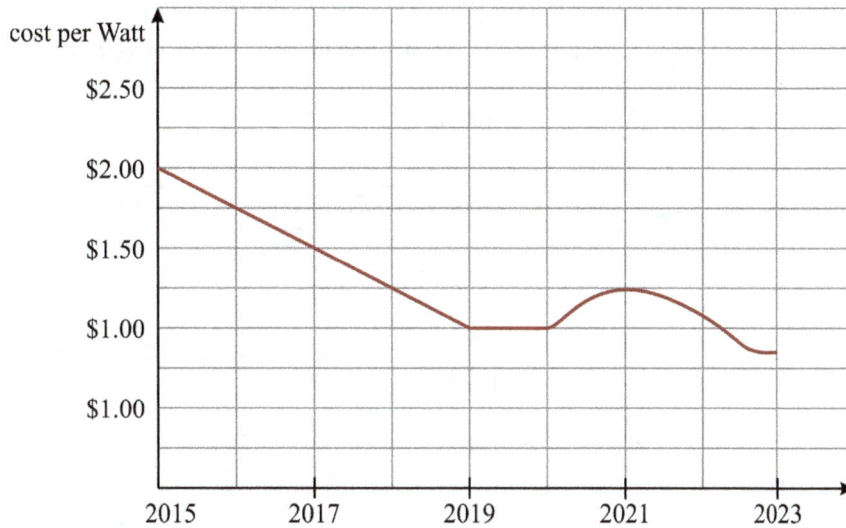

a. Find an interval where this function is decreasing in a linear manner.

b. Find an interval where the function was constant.

c. Describe the function from 2021 onward.

3. The graph shows the interest rate for savings accounts over time.

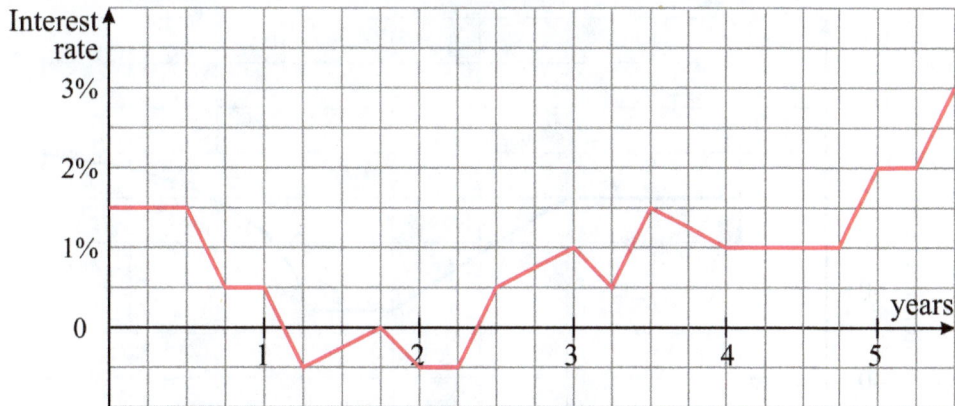

a. Find the longest interval where the interest rate was increasing.

b. Find the longest interval where the interest rate was constant.

c. What was the interest rate at 1.25 years?

d. Calculate the rate of change from 0 to 1 year, and also from 2 to 3 years.

4. Sketch a graph for a function that is...

- linear and increasing in the interval [0, 10]
- nonlinear and increasing in the interval [10, 20]
- constant in the interval [20, 25]
- decreasing in a linear fashion in the interval [25, 40].

Note: you need to design an appropriate scaling for the axes.

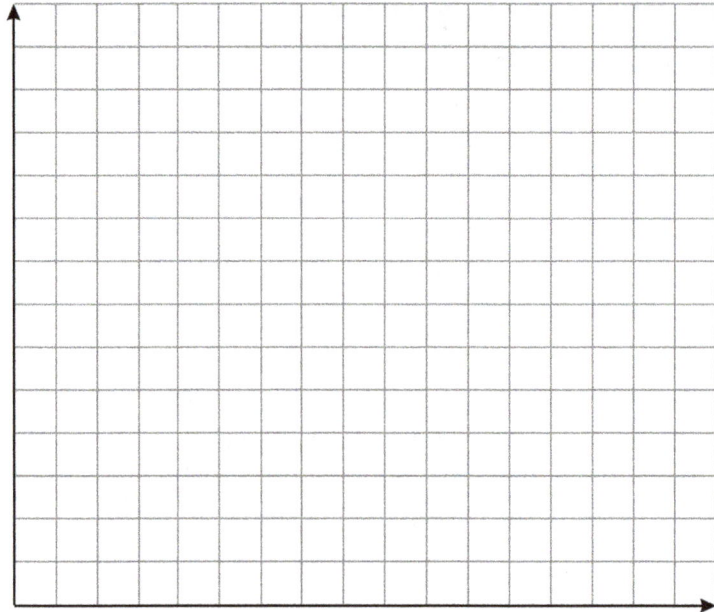

5. The graph shows the water level of a water reservoir over time. Make up a story of what is happening to affect the water level that matches the graph.

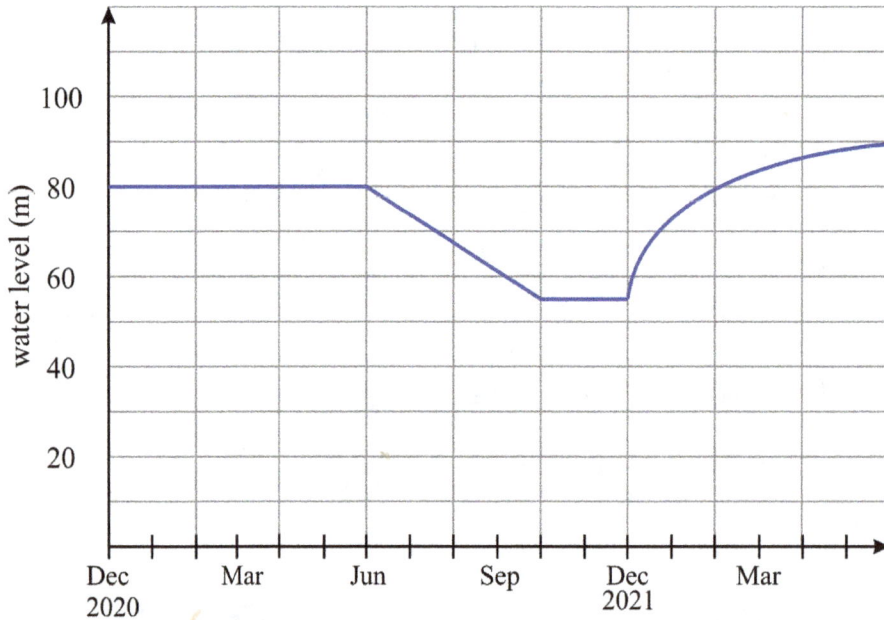

Describing Functions 2

Example 1. Tim is practising driving with a car that is parked in front of a wall. The graph shows the car's distance from the wall as a function of time. Can you imagine what is happening, based on the graph?

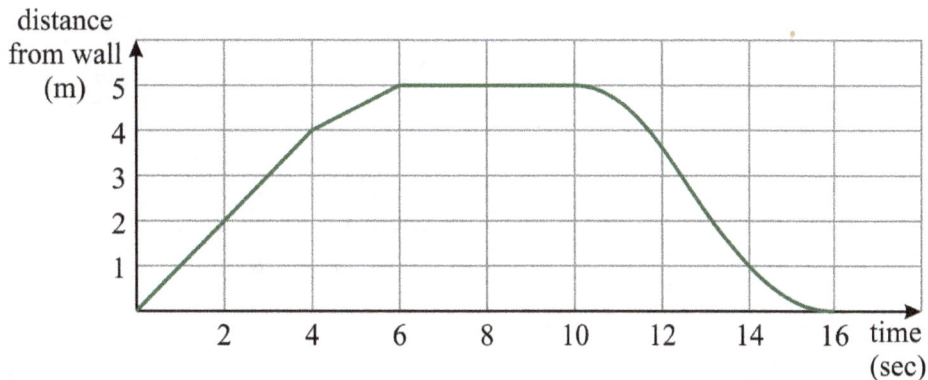

First, Tim backs off the wall with a constant (fairly slow) speed for four seconds. Then he slows down to a slower speed, still going away from the wall. At 6 seconds, he stops for four seconds, and doesn't move, the car being 5 m from the wall. Then, he starts towards the wall, in a smooth motion — first gradually gaining speed, then moving with a constant speed, then gradually with decreasing speed, until he is back to the wall in 16 seconds.

1. The graph shows the distance that Anne jogged and walked during a 1-hour walk, as a function of time.

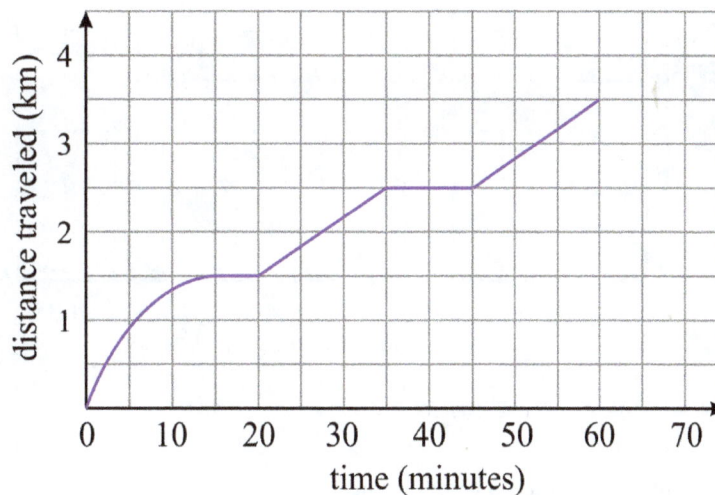

a. What was the total distance she covered in that hour?

b. What was her speed at 40 minutes?

c. At what intervals was she going at a steady speed?

d. Describe her speed in the first 15 minutes.

2. John plays with his dog Max at a local park. John stays in one place. The graph shows how far Max is from John. Make a story about Max's movements that matches the graph.

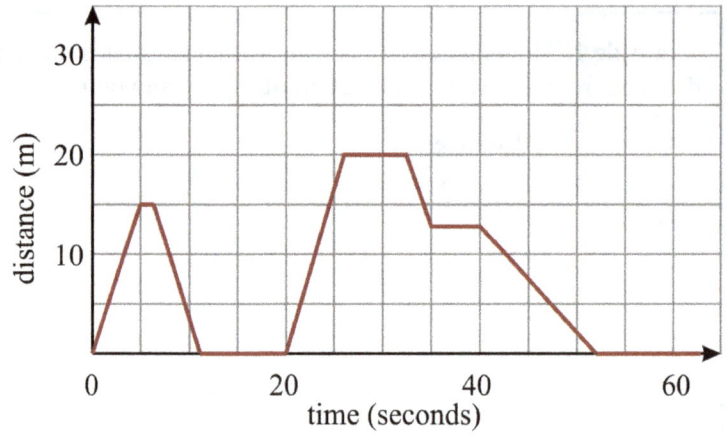

3. Draw a plot of a function that depicts the distance of Max, John's dog, from John, following the description. First, Max runs away from John for five seconds, until he is 20 metres away. Then Max slows down gradually during the next five seconds until he comes to a stop. Then he stops for 10 seconds, sniffing the ground. Then he still goes further away from John, for another ten seconds, but fairly slowly. Lastly Max turns suddenly and makes a mad dash back to John.

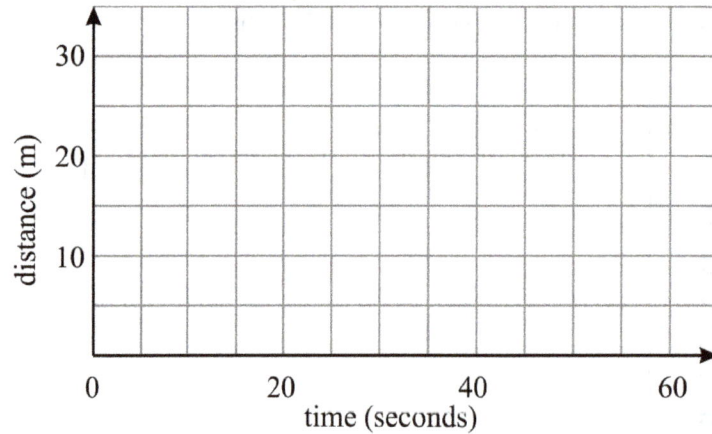

4. You throw a ball directly upwards. Sketch a graph depicting its height as a function of time.

First, the ball gains height quickly, but then it slows down. At 1.5 seconds, at a height of 11 metres, it stops gaining height and starts falling. First, it falls slowly, and then with increasing speed, until it reaches your hands again, at 3 seconds.

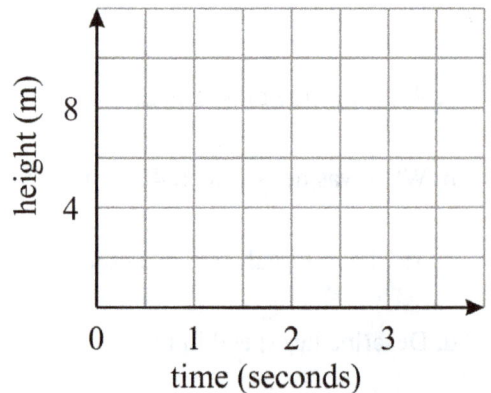

5. A rabbit and a turtle were racing. Make up a story of the race that matches the graphs. Be sure to specify which graph belongs to which animal. Who won?

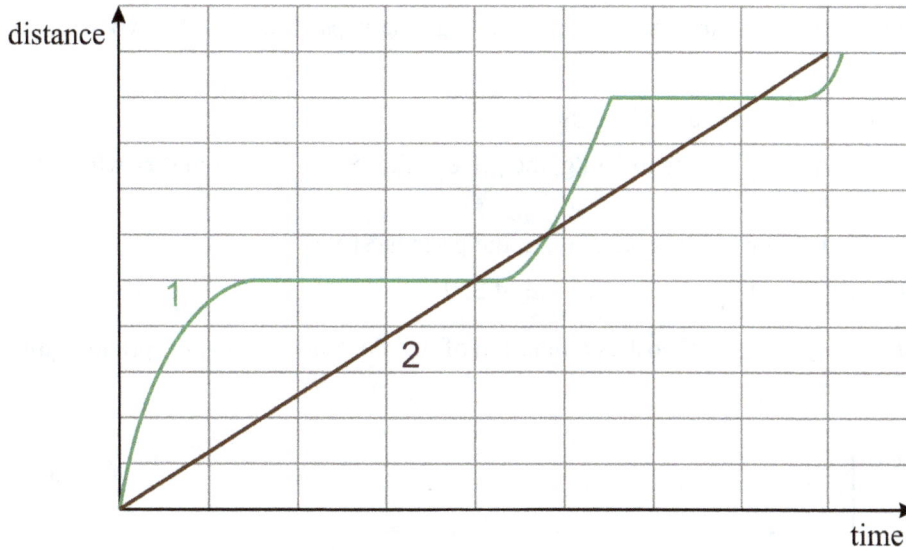

6. Plot a graph that matches the following story.

Jane fills an inflatable pool using a hose. The pool fills up at a constant rate, and after 10 minutes, there are 400 litres in the pool. Then Jane turns the hose off for 20 minutes. Then she continues filling it, and after 40 minutes more, the pool is full at 2000 litres.

Twenty minutes after that, something pokes a little hole in the pool, and it starts leaking at a slow rate, so that 1 hour later, it is 90% full.

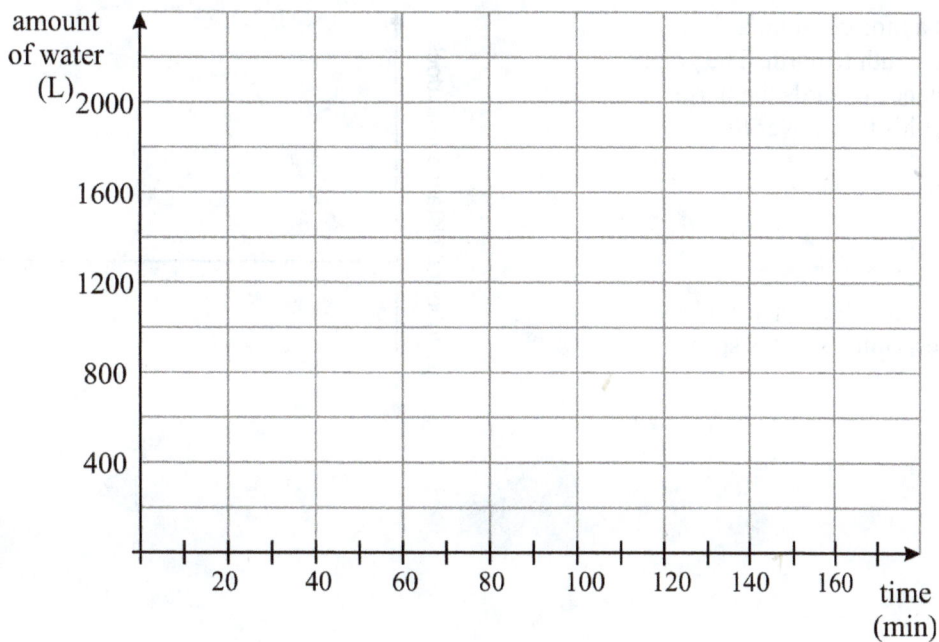

Describing Functions 3

1. Farmer Johnson has priced his strawberries in an odd way, perhaps. These are strawberries you pick yourself, at his farm.

   - If you pick up to 10 kg, the price is $1.50/kg.
   - If you pick more than 10 kg and up to 15 kg, the price is flat, $15. It won't matter whether you pick 11 kg, 12 kg or 15 kg — you pay $15.
   - If you pick more than 15 kg but less than 30 kg, the price is $1/kg.
   - If you pick 30 kg or more, the price is a flat fee of $30.

   Plot the function depicting the total cost as a function of weight. Choose an appropriate scaling for the vertical axis.

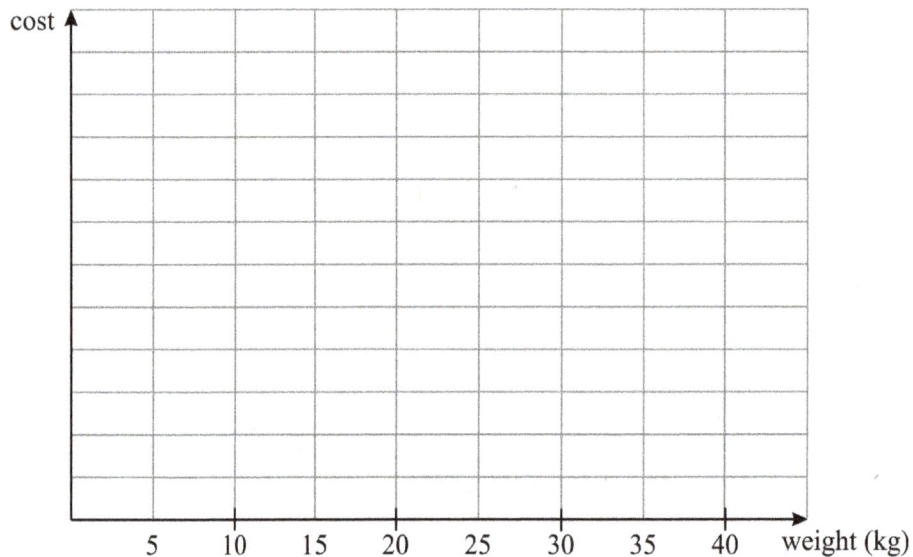

2. Mason drives a motorboat on a river that runs from south to north. The graph shows his distance towards the north from a dock at his town, over time.

   a. What happens at point A?

   b. Explain what an observer would see, including descriptions of his speed.

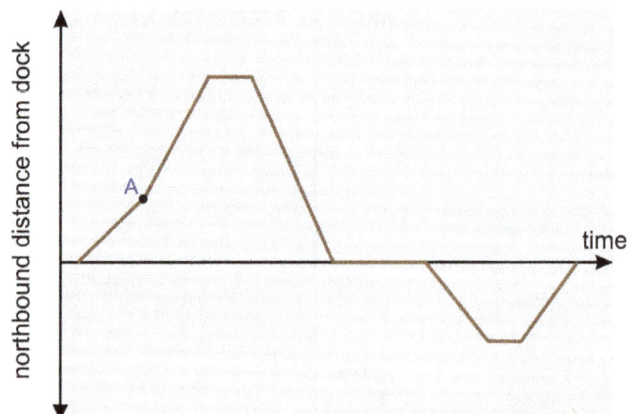

3. Graph the temperature as a function of time as described in this story. Note that you need to design a scaling for the vertical axis, and that the axis does not have to start at zero.

Story: At 6 AM in the morning, it was chilly, −4°C. The temperature gradually rose to −2°C by 8 AM. Then, from that time until noon, the temperature rose steadily to 4°C. Then it stayed there for three hours, and from 3 PM to 6 PM, it dropped steadily to −1°C.

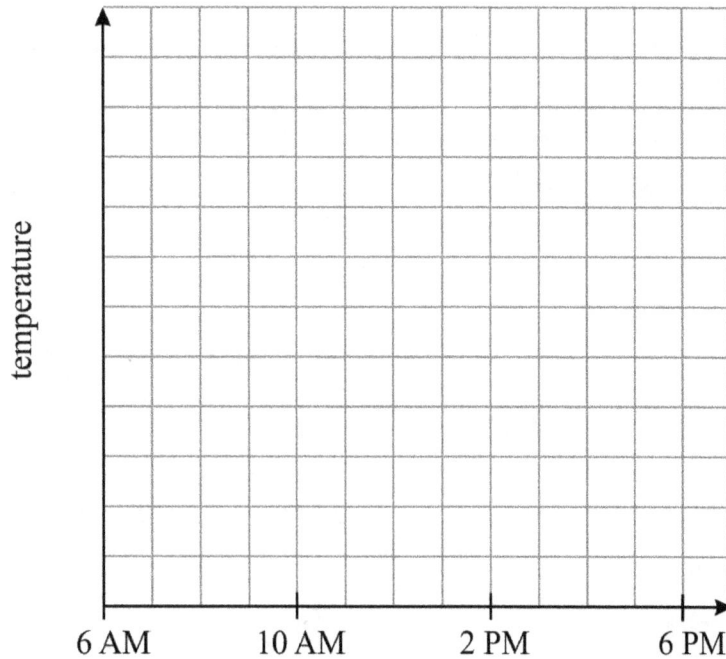

4. Janet is eating crackers out of a package. The graph shows the mass of the cracker package over time.

   a. What is happening during the time when the lines are horizontal?

   b. What happens at the vertical dashed lines?

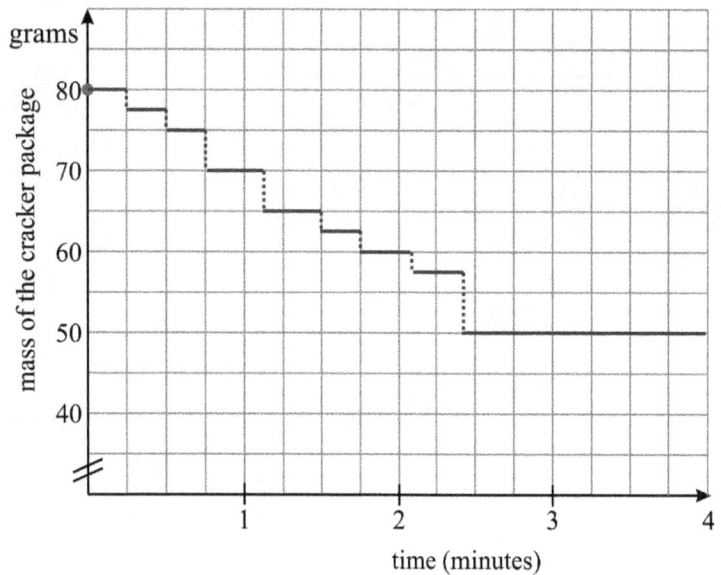

   c. Does she finish the package? How do you know?

   d. Based on the graph, make a guess for the mass of one cracker.

219

5. Match each description to a graph. Also, label the axes of each graph with the correct quantity (e.g. "price", "distance", "time"). Two graphs will not be matched.

   a. The price of chicken feed, where the price changes almost every month.

   b. The cumulative distance Alex runs on a jogging route, during an hour.

   c. The price of tomatoes is $3.50 per pound.

   d. The amount of water in a water container for cattle, over several days, where the farmer fills the container once it gets fairly low.

   e. The speed of a car on a shopping trip (a family drives to a store, shops, and comes back home).

   f. The amount of juice in a glass, over time, when it is first full, and Elsa drinks from it with a straw.

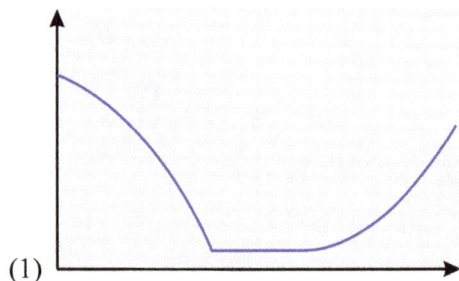

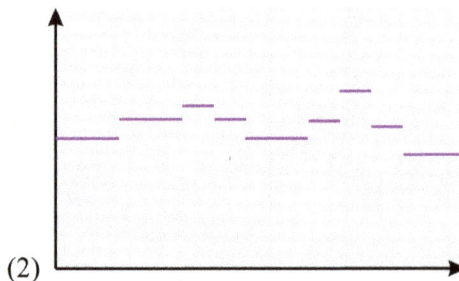

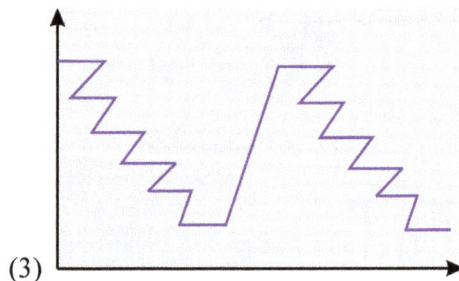

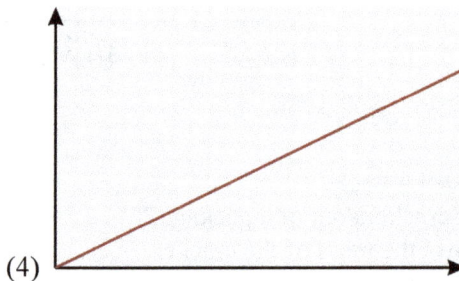

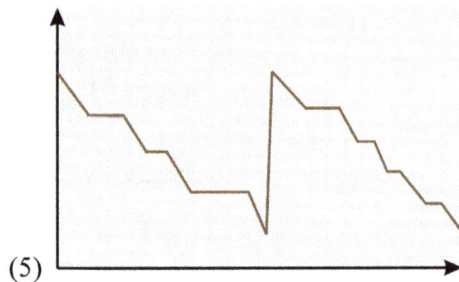

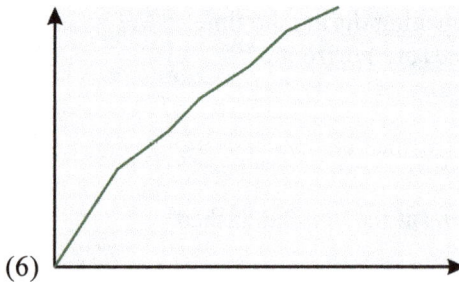

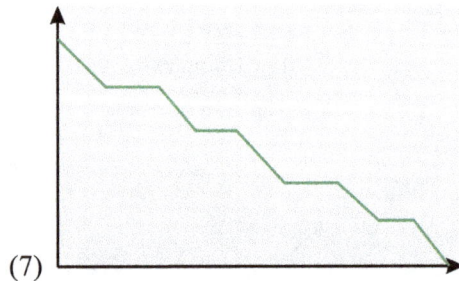

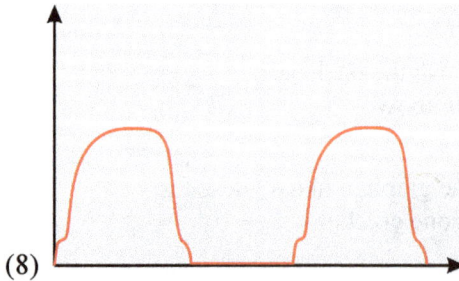

(1)

(2)

(3)

(4)

(5)

(6)

(7)

(8)

The first graph shows Carla's speed as she drives from home to town. Some parts of her journey have more traffic on the road than other parts, so her speed varies.

Use the speed graph to make a corresponding graph of the total distance Carla has driven.

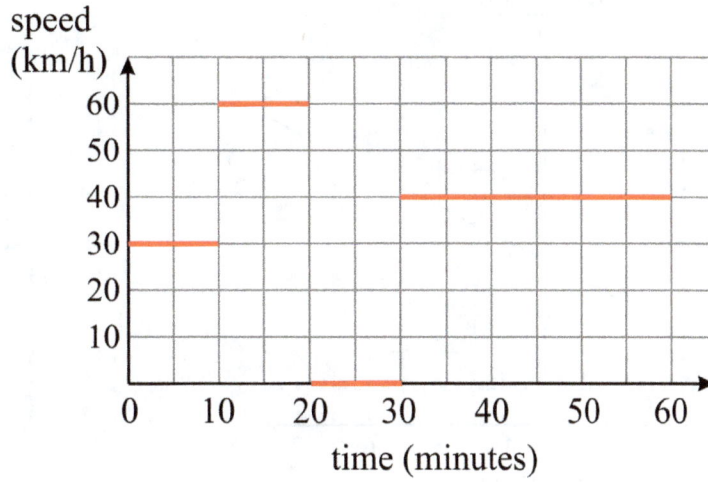

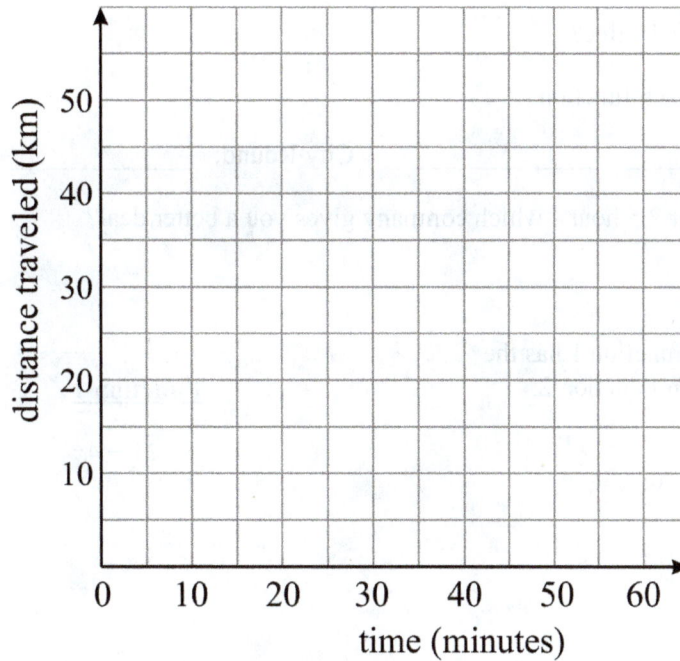

Comparing Functions 1

1. The graph and the table of values show the cost of renting a bicycle inside a city, for two different companies.

Mike's Bikes

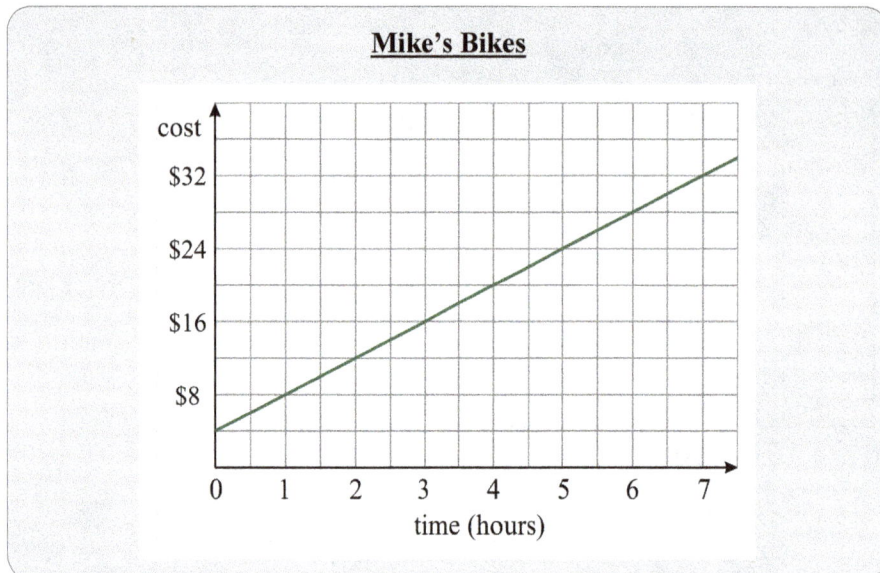

City Round

Time (hr)	Cost
0	$3.00
1	$7.50
2	$12.00
3	$16.50
4	$21.00
5	$25.50
6	$30.00

a. Which function has a greater rate of change?

b. Which has a larger initial value?

c. Write an equation for each function.

Mike's Bikes: _____ City Round: _____

d. If you rent a bicycle for 2.5 hours, which company gives you a better deal?

2. **a.** Samantha thinks that Function 1 has the greater initial value than Function 2, because 21 > 20.

  Is she correct? Why or why not?

Function 1

$y = 21 - 4x$

Function 2

x	y
1	20
2	17
3	14
4	11
5	8
6	5

b. Ethan feels that Function 1 has the greater rate of change, because each time x increases by one unit, y changes by four units, but for Function 2, y changes by only three units.

  Is he correct? Explain.

3. Richard is paying back two separate debts.

a. Which debt is he paying back quicker?

b. Which was a larger debt originally?

c. If he continues with the same rates, in how many months will he finish paying back debt 1?

Debt 2?

Debt 1

The loan payments are $350 monthly. After four months of payments, Richard has $2600 left to pay.

Debt 2

Time (months)	Amount of debt
0	$3400
1	$3150
2	$2900
3	$2650
4	$2400
5	$2150
6	$1900

4. Two airplanes, **Airplane 1**, and **Airplane 2**, fly for three hours.

The equation $d = 230t$ describes the distance Airplane 1 travels as a function of time (t).

The graph shows the distance Airplane 2 has travelled over time.

a. Which airplane flies with a greater speed from 0 to 1 hours?

b. Which airplane flies with a greater speed from 1 ½ to 2 ¼ hours?

c. Which covers a greater distance in 3 hours?

Airplane 1: $d = 230t$

Airplane 2:

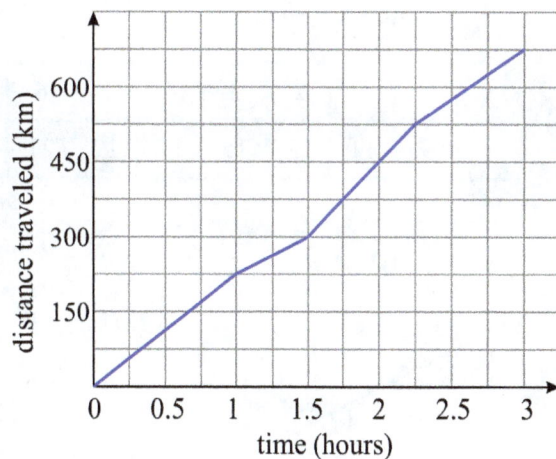

5. Sandra is comparing two international courier services. To send a package with **Service 1** costs you $15 plus $18 per half a kilogram. The fees for **Service 2** are listed in a table.

Weight (kg)	0.5	1	1.5	2	2.5	3	3.5	4
Cost ($)	57	73	89	105	121	137	153	169

   a. Which function has a greater rate of change?

   b. Which is a better deal if your package weighs 2.5 kg?

   c. Which is a better deal if your package weighs 6.0 kg?

   d. (*Challenge*) For which amount of weight are they equal?
      (*Hint:* use a graph, or an equation)

6. Below, you will see representations of six functions.

   a. Find two that depict the same function.

   b. Find two functions with the same initial value, but a different rate of change.

   c. Find two functions with the same rate of change, but a different initial value.

(1)

x	0	1	2	3	4	5	6
y	9	11	13	15	17	19	21

(2) $y = 42 - 3x$

(3)

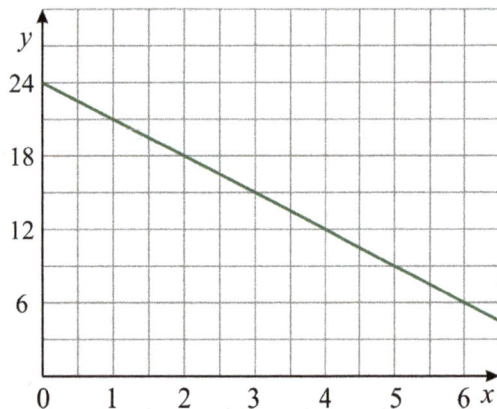

(4)

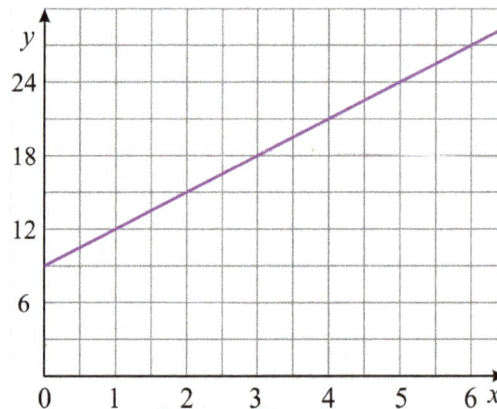

(5) From 4 pm till 8 PM, the water level at the harbor dropped steadily from 28 to 4 inches (part of low tide).

(6) $y = 3x + 9$

Comparing Functions 2

1. Which of the three functions represented below has the largest rate of change...

    a. in the x-interval [3, 4]?

    b. in the x-interval [4, 5]?

Function 1:

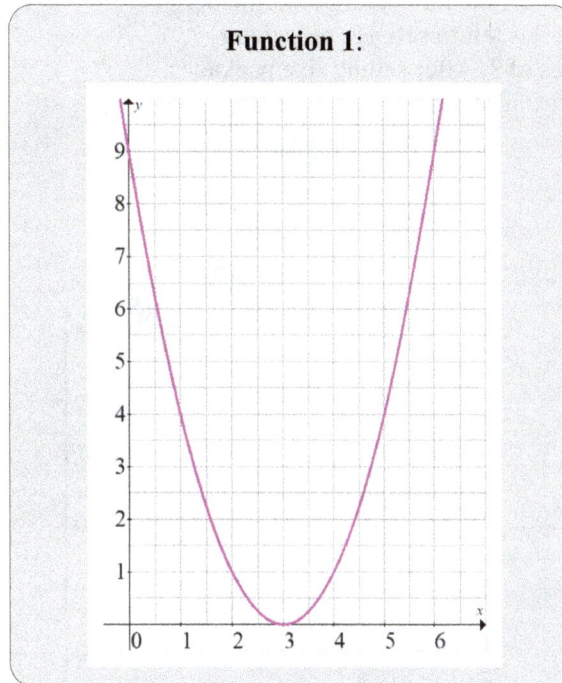

Function 2:

$$y = x - 3$$

Function 3:

x	y
1	2
2	5
3	7
4	10
5	12

2. Three functions are represented below.

    a. Which one has the largest initial value?

    b. Which one(s) are linear functions?

    c. Which one has the smallest rate of change in the x-interval [2, 8]?

    d. Describe each function in the x-interval [10, 12] as increasing, decreasing, or constant.

Function 1:

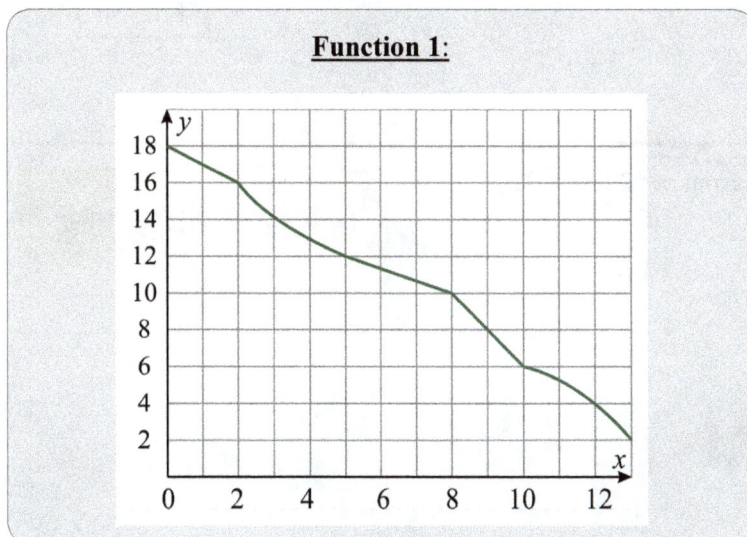

Function 2:

$$y = 20 + 1.5x$$

Function 3:

x	y
0	15
2	10
4	5
6	0
8	−5
10	−10
12	−15

3. Four different functions are each represented in two different ways. Find the matching pairs.

(1) Eric is tracking the number of home-made candles he has left to sell. He sells them in packages of 3. After selling five packages, he has 75 candles left.

(2) $90 - 5x$

(3)

x	y
4	9
8	14
12	19
16	24
20	29

(4)

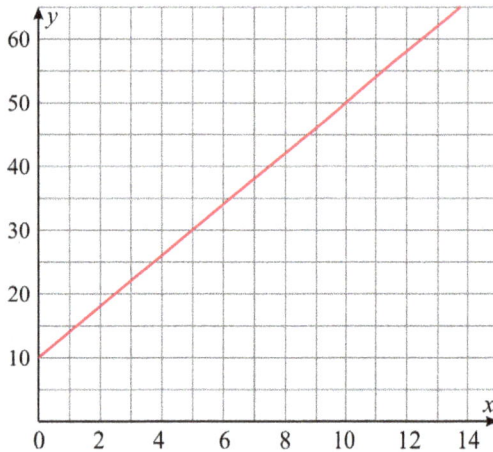

(5)

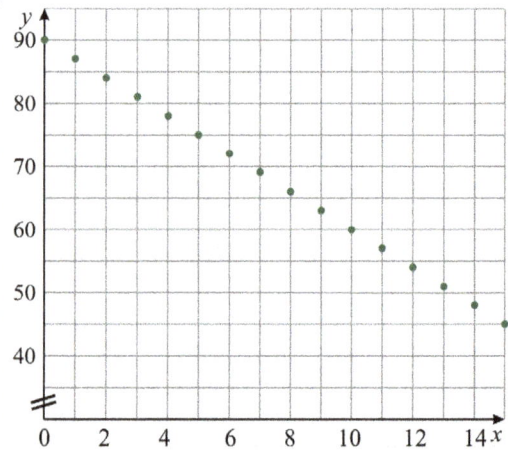

(6) $y = 4x + 10$

(7) Sandra sells muffins in her local neighbourhood. The cost of muffins is $1.25 per muffin, plus a $4 delivery fee.

(8)

x	y
0	90
3	75
6	60
9	45
12	30

Which function has the largest rate of change from $x = 2$ to $x = 3$?

Puzzle Corner

(1) $y = 2 - \dfrac{2}{x}$

(2) $y = \dfrac{1}{2}x + 2$

(3) $y = \dfrac{(x - 2)^3}{3}$

Mixed Review Chapter 4

1. Find the expressions that have the value 3^8.

a. $\dfrac{3^{16}}{3^2}$	b. 6^4	c. 24	d. $3^5 \cdot 3^3$	e. $\dfrac{3^{11}}{3^3}$	f. $\dfrac{3^7}{1/3}$

2. Write an equivalent expression using the exponent laws, without negative exponents.

a. $(b^2)^{-4} =$	b. $(-2y)^4 =$	c. $(7a)^{-2}$	d. $3x^2x^7y \cdot (-2)y^5 =$
e. $\dfrac{16x^9}{24x^3} =$	f. $\dfrac{5s^2}{s^{-3}} =$	g. $\left(\dfrac{4x}{-6}\right)^2 =$	h. $\left(\dfrac{a}{2b^2}\right)^4 =$

3. Solve. Give your answer as a normal number and also in scientific notation.

a. $5 \cdot 10^{-4} + 2 \cdot 10^{-3}$	b. $9 \cdot 10^6 + 2 \cdot 10^7$
c. $5 \cdot 10^{-2} - 8 \cdot 10^{-3}$	d. $8 \cdot 10^6 - 7 \cdot 10^5$

4. Let $a = 8 \cdot 10^9$ and $b = 2 \cdot 10^7$. Find the value of (a) their product; (b) their quotient. Give your answers in scientific notation.

5. **a.** The volume of a cone is V $= \dfrac{A_b h}{3}$, where A_b is the area of the base and h is the height of the cone. Solve this for h.

  b. What is the height of a circular cone with a volume of 20 900 cm^3 and a bottom radius of 25.0 cm?

6. Draw dilations.

a. Draw a dilation of kite DEFG from point E, with scale factor 1/2.	b. Draw a dilation of triangle ABC from point C, with scale factor 3.

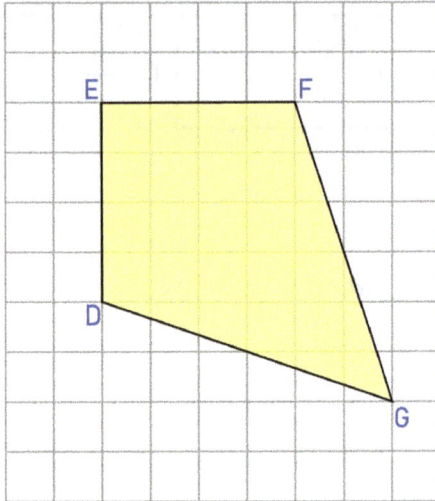

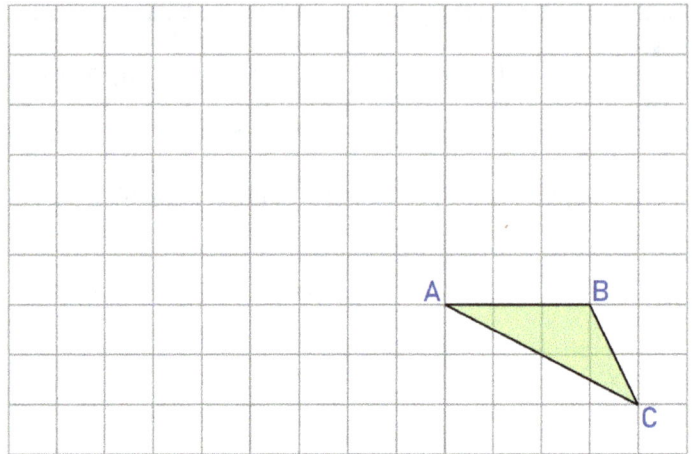

7. The price of gasoline went from 182.5 cents to 189.2 cents per litre. What was the percent of increase?

8. An online curriculum that normally sells for $72.95 is on sale for 30% off. A sales tax is added to your total, and you pay $53.92. What is the sales tax percentage?

9. Solve. Note carefully how the two equations (a) and (b) differ, and how that affects the solution process.

a. $$3x + \frac{x+3}{5} = 1$$

b. $$3x - \frac{x+3}{5} = 1$$

10. Tell, without fully solving the equations, whether each equation has one unique solution, no solution, or an infinite number of solutions.

    a. $y = 6 - 7y$

    b. $6 - 7y = 2 - 7y$

    c. $-7y + 14 = 7(2 - y)$

    d. $-7y - 2 = -2$

11. Solve.

a. $10 + 3(a + 5) \;=\; 2(a - 6) - 4a$	**b.** $20x - 2(x + 1) \;=\; 10 - (x - 5)$
c. $\frac{1}{6}x - 1 \;=\; 1 + \frac{4}{5}x$	**d.** $2z + \frac{2}{5} \;=\; \frac{1}{4}z - 1$

Chapter 4 Review

1. Change some thing(s) in this table so it represents a function.

Input	Output
Name	**Age**
Fifi	2
Bella	5
Max	
Luna	2
Charlie	6
Luna	3

2. Why is the relationship depicted by the graph *not* a function?

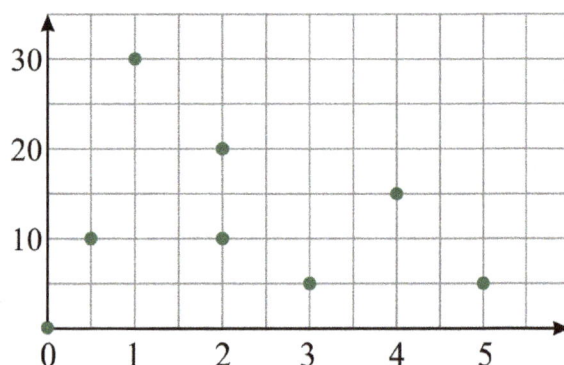

3. Describe this function by intervals where it is increasing, decreasing, or constant. Include also whether it is linear or nonlinear in those intervals.

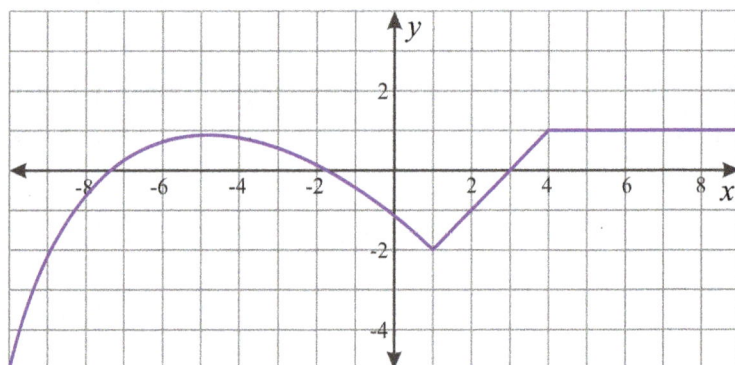

4. Give an example where the cost of something is a function of time, and where the function is *not* linear. Use a table of values to give the function.

Time (hours)	0	1	2	3	4	5	6	7	8	9	10	11	12
Cost ($)													

5. The graph shows the value of a car over time.

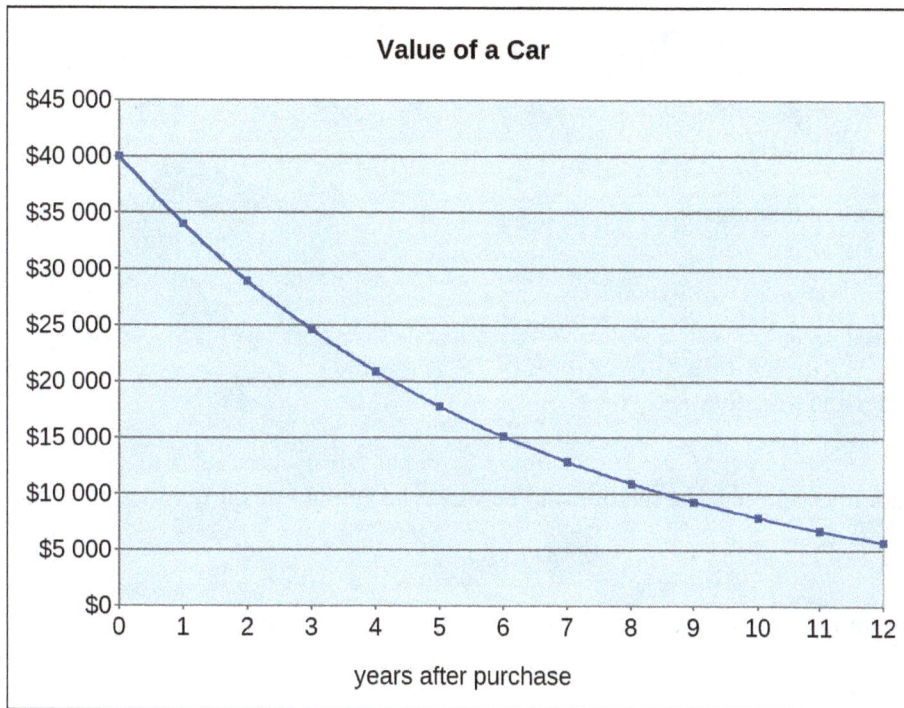

Value of a Car

a. Is this a linear function?
   How can you tell?

Use approximate values that you can read from the graph, and find the rate of change:

b. from 1 to 2 years after purchase

c. from 6 to 7 years after purchase

6. Jayden and his sister race on bicycles from school back home, a route that is 2.4 km long.
   The equation $d = 0.35t$ represents the distance (d, in km) that Jayden has ridden, as a function of time
   (t, in minutes). The table below shows the distance his sister has bicycled, at various points in time.

Function 1 — Jayden:

$d = 0.35t$

Function 2 — his sister:

time (minutes)	0	1	2	3	4	5	6
Distance (km)	0	0.32	0.72	1.1	1.44	1.79	1.98

a. Which function has a greater rate of change from $t = 2$ to $t = 4$ minutes?
   What does that represent in terms of real life?

b. Classify each function as either linear or nonlinear.

c. Assume his sister continues with the same speed till the end as what
   she is riding between 5 and 6 minutes. Who will reach home first?

7. The table below shows the depth of the snow as a function of time from the beginning of a blizzard.

time (hours)	0	1	2	3	4	5	6
depth (cm)	43	50	57	64	71	78	85

a. What is the rate of change?

What does it mean in this situation?

b. What is the initial value?

What does it mean in this situation?

c. Write an equation to represent the relationship between the amount of snow and time in hours.

d. If the blizzard started at 2:30 PM, at what time was the snow 65 cm deep?

e. How deep will the snow be at 9 PM?

f. Plot the equation you wrote in (c).

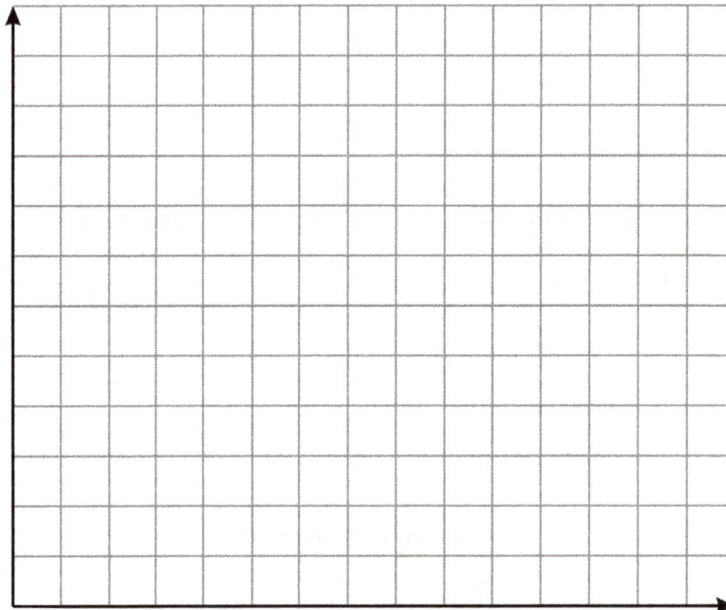

8. Maria is in math class (in a classroom). The graph shows the distance between her and the blackboard, as a function of time. Make up a story that matches the graph.

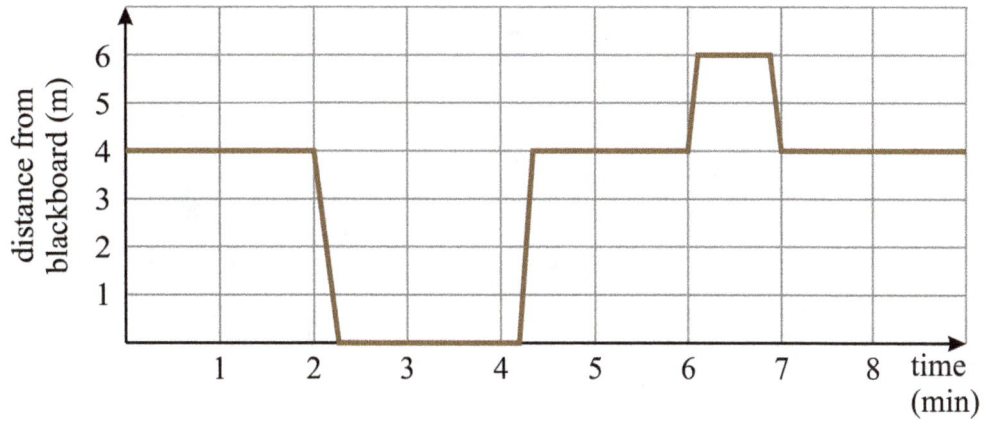

9. Greg is running along his usual running route, and his son Trevor goes with him riding a bicycle, but they don't exactly stay together. The two graphs show the total distance each has covered, as a function of time.

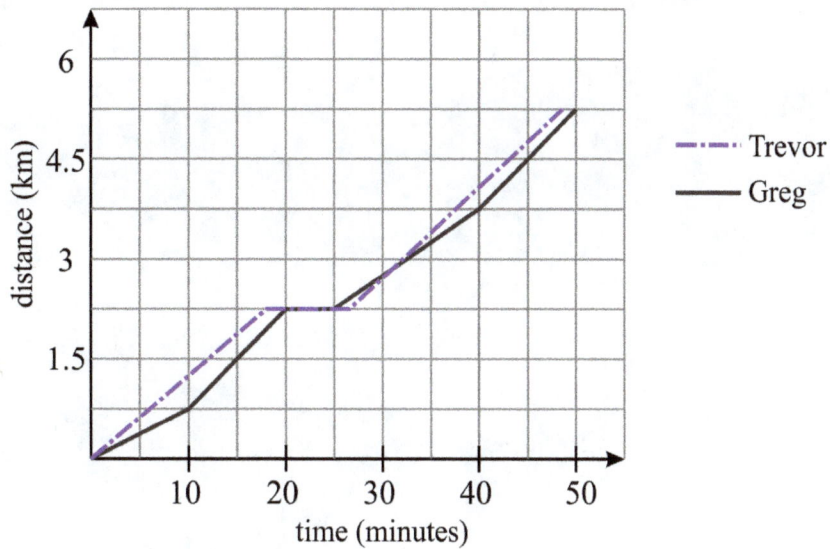

Trevor

Greg

a. From 10-20 minutes, who is going faster?

b. Find two points in time where the two meet. What distance have they travelled at those times?

c. Who finishes the route first? About how much quicker than the other (estimate from the graph)?

www.ingramcontent.com/pod-product-compliance
Lightning Source LLC
Chambersburg PA
CBHW080532220326
41599CB00032B/6284